PERGAMON INTERNATIONAL LIBRARY
of Science, Technology, Engineering and Social Studies
The 1000-volume original paperback library in aid of education, industrial training and the enjoyment of leisure
Publisher: Robert Maxwell, M.C.

RESOURCES SOCIETY AND THE FUTURE

Other Titles of Interest

GABOR *et al.*	Beyond the Age of Waste
GVISHIANI	Science, Technology and Global Problems
MAROIS	Towards a Plan of Actions for Mankind (5 vols.)
PECCEI	The Human Quality
SECRETARIAT FOR FUTURES STUDIES	Solar Versus Nuclear: Choosing Energy Futures
SIMEONS	Coal: Its Role in Tomorrow's Technology
UNECE	Coal — 1985 and Beyond
VELIKHOV	Science, Technology and the Future

RESOURCES SOCIETY AND THE FUTURE

A Report Prepared for the Swedish Secretariat for Futures Studies by

TOMAS BERTELMAN, ERNST HOLLANDER, CARL-AXEL OLSSON,
MARIANNE PARMSUND, STAFFAN SOHLMAN (project leader) and UNO SVEDIN

Translated from the Swedish by
ROGER G. TANNER

PERGAMON PRESS
OXFORD · NEW YORK · TORONTO · SYDNEY · PARIS · FRANKFURT

U.K.	Pergamon Press Ltd., Headington Hill Hall, Oxford OX3 0BW, England
U.S.A.	Pergamon Press Inc., Maxwell House, Fairview Park, Elmsford, New York 10523, U.S.A.
CANADA	Pergamon of Canada, Suite 104, 150 Consumers Road, Willowdale, Ontario M2J 1P9, Canada
AUSTRALIA	Pergamon Press (Aust.) Pty. Ltd., P.O. Box 544, Potts Point, N.S.W. 2011, Australia
FRANCE	Pergamon Press SARL, 24 rue des Ecoles, 75240 Paris, Cedex 05, France
FEDERAL REPUBLIC OF GERMANY	Pergamon Press GmbH, 6242 Kronberg-Taunus, Pferdstrasse 1, Federal Republic of Germany

First edition 1980

British Library Cataloguing in Publication Data

Swedish Secretariat for Futures Studies
Resources, society and the future. - (Pergamon international library).

1. Natural resources - Sweden
2. Natural resources - Case studies

I. Title II. Sohlman, Staffan

333.7'09485 HC373.5 79-40724

ISBN 0-08-023266-3 Hardcover
ISBN 0-08-023267-1 Flexicover

In order to make this volume available as economically and as rapidly as possible the typescript has been reproduced in its original form. This method has its typographical limitations but it is hoped that they in no way distract the reader.

Printed and bound in Great Britain by William Clowes (Beccles) Limited, Beccles and London

Contents

Foreword

Questions concerning resources have attracted an increasing interest during the 1970s. A succession of studies have drawn attention to possible clashes between environmental interests and raw materials, between the development aspirations of the industrialized and developing countries, between short-term desires and long-term possibilities. Simplifying matters somewhat, one can distinguish between two debates in Sweden, one of them centering round national interests, i.e. Sweden's needs for industrial production and exports, and the other concerning long-term questions of human survival, viz the environment, the resource base, and questions of equity between industrialized and developing countries.

In the forecasting study entitled Resources and Raw Materials: Sweden's Long-Term Options in an International Perspective, our aim has been to link these two perspectives. In the present final report of this project, the project group has tried to summarize its thoughts concerning developments in the fields of resources and raw materials during the remainder of this century. Of course, we had hoped to be able to say more and to say it more definitively. Most of the work in this study has been devoted to compiling the series of interim and special reports listed in Appendix 2. In many cases the text of this final report is based on more exhaustive arguments to be found in the interim studies.

This final report has evolved in the course of continuous discussions within the group. On the whole, it is a collective product, even though individual members have raised certain objections or would have preferred more far-reaching conclusions on certain points.

There are many whom we would like to thank for the assistance they gave in connection with this study - too many people, in fact, to be mentioned by name in this brief foreword. One of the most encouraging features of our work has been the spontaneous interest and helpfulness with which we have met, not least on the part of the inter-departmental reference group which has supported us throughout our work.

Stockholm, September 1977

Tomas Bertelman
Ernst Hollander
Carl-Axel Olsson
Marianne Parmsund
Staffan Sohlman (project leader)
Uno Svedin

Preface

By decision of June 30th, 1971, the Prime Minister appointed a Working Party under the chairmanship of Mrs Alva Myrdal, Cabinet Minister, to deal with questions of future studies in Sweden.

The report of the Working Party, "To Choose a Future - a basis for discussion and deliberations on future studies in Sweden", was submitted to the Prime Minister on August 25, 1972.

Following the proposal of the Working group a secretariat for futures studies was set up in 1973 by the Swedish government. The present book - Resources, Society and the Future - is the final report of the first completed futures studies project. The perspective for these studies was explained in the introduction by Mrs Alva Myrdal on the occasion of publishing an English version of the report "To choose a future", and is still valid:

"The concept of the welfare state has hitherto been seen as closely harnessed to the rapid development of science and technology. However, this development is now also being recognized as a source of danger for our societies because of over-hasty transformations and unforeseen negative side effects. There is broad agreement that scientific and technological capacity will be a crucial factor in our common efforts to bring acceptable material standards of living to all people. But are the present large-scale technologies really furthering this aim? What alternative choices exist? How are cumulative technological innovatory processes to be reconciled with slower changes in social values and institutions? These are the questions to which we ought to devote more attention, if the total development efforts within and between countries are to bring us nearer the all-embracing goal of an improved quality of life for all mankind.

A less naive approach towards the role of science and technology in the promotion of development is emerging. This holds for rich and poor countries alike. An international debate that has been going on for several years has stressed the importance of determined efforts to analyze and "create" the future. The present tendency to know more and more about less and less creates problems for society as a whole; the era of extreme specialization must come to an end.

"To Choose a Future" was produced and published as a contribution to a national and international debate on future problems. The chief aim of the report was to create in Sweden a framework for a continous dialogue between the general public

and their political representatives, the scientific world and interest groups of all kinds. International development in science and technology has been dominated by heavy military and commercial interests. Hence, it is not surprising that most future studies produced so far have also been sponsored by military establishments and by the world's major multinational corporations. Such studies should be the concern of all countries. Critical awareness is needed, as these studies may be based on scales of value that are not democratically acceptable. We must avoid any "colonizing of the future" by powerful interest groups, national or international.

One main idea behind the report is the need for public participation in the work on future studies. We must always devote much energy to making complex problems commonly understandable. Future studies and the discussion about different futures must not be left to a new breed of specialists or to an elite who claim to know what is best for everybody. The democratic control of this work must never weaken, and public participation in vital long-range decisions must be safeguarded and deepened.

International events during the last year have underscored the urgency of the pair of ideas-future-orientation and internationalism - that are basic in "To Choose a Future". I have, however, learned with great satisfaction that international organisations, such as UNESCO and UNITAR, have decided to establish and promote activity in the field of future studies. It is my sincere hope that the international community will extend its responsibility - also in practical terms - to include future generations".

1 Motives for a Futures Study of Resources and Raw Materials

Introduction

Interest in questions concerning resources and raw materials has risen steeply in recent years. Many warnings have been issued on the exhaustion and mismanagement of the world's endowments of minerals, water, soil and air. Practically every debate focussing on the future incorporates quite gloomy pronouncements concerning the "carrying capacity" of the earth, in its various forms.

Warnings of an approaching ecological crisis have come from a variety of political vantage points. The list of problems can be roughly summarized as follows.

1. A growing world population means rising material needs, and tremendous growth will therefore be required in order to more fully satisfy the needs of this population.

2. Most of the energy hitherto employed for industrial processes comes from non-renewable sources. Assuming exponential growth, the energy sources which could be available for exploitation will not last for more than another century or two at the most. It is theoretically conceivable that new sources of energy will replace the old ones, but it is hard to predict when the proposed ideas (particularly fusion energy) will be translated into practice.

3. Industrial processes also require metals and industrial minerals. Exponential growth will lead to the exhaustion of many of these minerals, even if reserves should prove to be several times greater than is currently supposed.

4. Water consumption in industrial processes has reached a level where supplies of water from the natural cycle are no longer sufficient. In many cases this has led to over-exploitation of groundwater resources, a course of action which is bound to disrupt the hydrological cycle and produce climatic changes.

5. The scope for an increase in food production is limited by the fact that more than half the world's arable land is already under cultivation and yield per acre cannot be raised indefinitely.

Above a certain limit increases in yield per acre result in ecological imbalance (erosion, toxicity, a decline in the number of species).

6. Pollution - or rather, disruptions of the equilibrium in the "metabolism" existing between nature and human society - is a side-effect of the industrial process and increases with industrial growth. The long-term effects are often impossible to predict or monitor.

7. The mental strains resulting from the rapid growth of our urban agglomerations, with its concomitant rootlessness, occupational stress, monotony, etc., is even more difficult to quantify or control with the aid of standard policy instruments.

8. The consequences of "thermal destruction of the environment" i.e. the threat to the earth's thermal equilibrium, are also hard to foretell.(1)

However, these various prophecies of doom have not remained uncontradicted. For example, their authors have been criticized for overlooking control mechanisms that are built into human societies. Thus, population growth seems to tail off with rising living standards. Scarce resources can be replaced with resources that are more abundant. Resources cannot be viewed as a fixed entity; they are dependent on the level of technology and on social organization. Economists have frequently contended that the best safeguard against the depletion of resources is furnished by the pricing system of the market economy, which provides built-in opportunities for substitution and continuing technological progress. This view is represented by the following extract from an issue of the Swedish journal Ekonomisk Debatt.(2)

> "Today's economists have learnt their lesson and can see abundant and practically inexhaustible possibilities of technological development, new deposits, the replacement of scarce raw materials and shifts in the pattern of consumption..... Our peculiar role in the current debate is due to the prophecies of doom being based on theories which we find dubious and indeed ridiculous".

Another characteristic of the debate on resources is that, roughly speaking, it proceeds at two levels. The forecasts and conferences of the international organizations on the subjects of population, raw materials, food supply and so on are principally aimed at achieving political solutions, often of a short-term nature. Proposals concerning various forms of assistance and the stabilization of raw material prices are examples of solutions of this kind. This debate does not really challenge the world economic structure or politico-economic systems.

The debate conducted on the other level raises questions that are more wide-ranging. Here resources are discussed in terms of "questions of human destiny", and quite often demands are made - on ecological, political or ethical grounds - for a complete

(1) This enumeration is a rough precis of the points raised by Hans-Magnus Enzensberger in an essay entitled "Till kritiken av den politiska ekologin" (To the critique of political ecology)

(2) Nils Lundgren, Ekonomisk Debatt 8, 1974, p.446.

transformation of the social and global order.

Of course, the fact that interest in questions concerning resources and raw materials has grown of late need not necessarily imply any decisive deepening or transformation of the real problems in these sectors. Nor, as we shall see presently, is this the first time that fears have been expressed concerning the depletion of resources. It is not inappropriate, therefore, to try to pinpoint reasons for the interest which has come to centre on resources and raw materials during the 1970s. In doing so we shall at the same time be able to give some of the reasons justifying a Swedish futures study on the subject of resources and raw materials.

We may begin by referring to the futures studies produced for the Club of Rome and entitled "The Limits to Growth" (1972) by Dennis L. Meadows and "Mankind at the Turning Point - the Second Report to the Club of Rome" (1974) by Mihajlo Mesarovic and Eduard Pestel. These studies help to explain the interest focussing on resources, because they forecasted that, within a none too distant future, the world's natural resources would impose limits on continuing economic growth. But if the reports from the Club of Rome did so much to draw attention to the question of resources and raw materials, our next step should be to find out what induced the Club of Rome to interest itself in these problems. One cannot discount the possibility that futures studies of this kind are symptomatic of more deep-rooted problems in the economies of the developed industrialized countries which were later to present a number of effects. While the studies undeniably put life into the discussion, a number of other conceivable causes for concern at a later stage will be considered here, without any claim to exhaustivemess.

On the more factual level, we may recall that the raw materials boom which occurred between 1972 and 1974 aroused fears, especially in the industrialized countries, of an impending drastic reversal of the price relationship between raw materials and finished products as it had evolved up till then.

Many people interpreted the price rise as indicating a long-term tendency towards a growing shortage of raw materials, both globally and regionally. The long row of national surveys concerning supplies of natural resources which followed in the wake of the raw materials boom was a clear sympton of this growing concern. Quite often these surveys were intended as a documentary input for a national raw materials policy.(1)

The successful action undertaken in the autumn of 1973 by the petroleum exporting countries also played an important part. Oil prices quadrupled. This event seemed to suggest that those developing countries which export raw materials were in a position, thanks to their possession of raw material resources, to exert political and economic pressure and thereby achieve lasting changes in raw materials markets.

A long-term transition from a buyers' to a sellers' market seemed certain. At the same time, of course, one cannot discount the possibility that "The Limits of Growth", widely publicized as it was helped to heighten awareness of the 'power' implied by the possession of raw materials.

Developments in raw materials markets during the past few years have greatly

(1) For a survey of various studies of this kind, see "Resurser och ravaror - nagra utgangspunkter for en framtidsstudie" (Resources and raw materials - some points of departure for a future study), interim report no. 301 of the RoR Study 1975, p. 119 ff.

increased the interest taken in various forms of intervention in the world market. This interest has been evidenced by producer countries and producers' associations as well as by consumer countries, but there is a great deal of disagreement concerning the measures which ought to be taken. The bad harvests of 1972 and the price rise and supply problems to which they gave rise are examples of a contributory cause. We shall be returning to this subject in due course.

Questions concerning raw materials have also come to be discussed more actively within such major international bodies as the UN, GATT and UNCTAD. Instead of complete review of these activities, a few examples will suffice.

The UN World Food Conference of 1974 came to centre around questions related to internationally co-ordinated grain storage. A special body - the World Food Council - was set up to ensure that stocks were organized on this basis.

Raw materials questions formed part of the agenda of the Sixth Special Session of the UN General Assembly in 1974. That Session adopted the principle that every state should have full sovereignty over its own natural resources, and it also adopted a statement in support of producer co-operation in the raw material sector, though most of the industrialized countries lodged reservations on the latter point.

Raw materials questions have been in the forefront of UNCTAD's proceedings ever since the first conference, in 1964. Over the years many proposals have been put forward with the aim of reversing the deterioration in the prices received by developing countries for their raw materials. No very extensive policy measures have resulted, but the discussion started during UNCTAD IV was pursued further between 1975 and 1977 in the Conference on International Economic Cooperation (CIEC).

Background to the study of resources and raw materials

The "Resources and Raw Materials" project set up by the Swedish State Secretariat for Futures Studies has to be viewed against this background. It was no coincidence that, when the problem areas to be covered by the Secretariat were being discussed, elucidation of 'the question of resources' was found to be urgently needed. (Appendix 1)

Now that the final report of the project is to be presented, it is pertinent to ask what we have been able to accomplish in this very wide problem area. Questions relating to the future are concerned with many other things besides resources. At the same time, questions concerning resources can hardly be viewed in isolation from other major issues connected with the future. In this study, moreover, we have only been able to consider certain areas of the problem of resources, as the reader will soon find out. We have tried to justify our priorities as extensively as possible.

One important restriction from the outset was that our study was not to include the energy question, because a special energy study was started at the same time as the resource study.(1) Of course, this does not mean that energy has been excluded altogether; it has been taken into consideration insofar as it is relevant to raw materials and their extraction.

(1) The energy study group published a final report (in Swedish) during 1978 called "Sun or uranium - to choose an energy future". It will be published in English as well.

The debate on raw materials involves a variety of perspectives. The concrete realities of the grain harvest in India are founded on the question of survival. The loading and unloading of cargoes in the London docks, for example, is directly subject to the economic cyclical problems of the advanced industrial society. (Pictures: Goran Bergsten and Press Association Photos).

The RoR (Resources and Raw Materials) study was not governed by any express terms of reference. It was stated in the programme that the study was to focus on Sweden's supplies of raw materials, viewed in a long-term perspective and against an international background. This has given rise to a difficult problem of balance. Should the emphasis of our work be on Sweden's supplies of raw materials, or should its main concern be with international solidarity and with Sweden's possibilities of pursuing a raw materials policy which actively serves that purpose? Even though we made no explicit choice, our interest came to focus at an early stage on global problems and on political and economic conditions with a bearing on resource problems.

Another problem which this study encountered has been the uncertainty as to which group or groups the study ought primarily to be aimed at. One of the Secretariat's declared aims for its futures studies is to have two sets of recipients. They are meant to furnish decision-makers and politicians with documentation, but they are also expected to help widen the debate on future issues between groups and organizations and among citizens generally.

These various alternatives are not mutually exclusive, but we cannot deny that 'politically relevant issues' have sometimes been given priority over activities of a more 'undisciplined' nature. We have e.g. not presented any far-reaching Utopian arguments, and we have for the most part refrained from recommending a particular radical alternative.

This may be viewed as a shortcoming, and it may lead some readers to feel that the results of the study are less exciting than they might have been. A main reason why we have been forced to focus on 'realistic' problems rather than those of a more 'Utopian' nature is the effort needed to penetrate the vast field of factual problems presented by resources and raw materials. Most of the interim studies published in the course of the project(1) have simply been reviews of the state of knowledge concerning various issues within the wider complex of problems. (This however has not prevented the interim studies from being directed towards the future debate). Predictions based on value premises different from those current today can be rendered more accurate by the possession of sound facts concerning the present situation and historical developments. In addition to factual documentation, it has been our aim to present something of a catalogue of arguments which, we hope, will make it possible for the reader himself to construct alternative pictures of the future and to undertake a better appraisal of those presented elsewhere.

The membership of the project group has been such that the social sciences have outweighed the natural sciences. It is quite possible to criticize both the interim studies and the present final report for a limited treatment of the ecological and environmental aspects of the problem, despite the efforts that have been made to give these aspects the best possible coverage. On the other hand it seems perfectly reasonable to assume that developments during the next few decades will continue to be dominated by powerful political and economic tendencies.

Figure 1 comes from the first interim study, "Resources and Raw Materials - Some Points of Departure for a Future Study". In view of the reservations which have already been made, it goes without saying that the lofty aspirations of the study have not been fully realized. As figure 1 indicates, the intention was to give considerable scope to work on pictures of the future, otherwise termed "scenarios".

(1) Appendix 2 where a list of titles are given. Appendix 3 contains examples of sumaries of three of the interim studies.

Figure 1. Development of the study

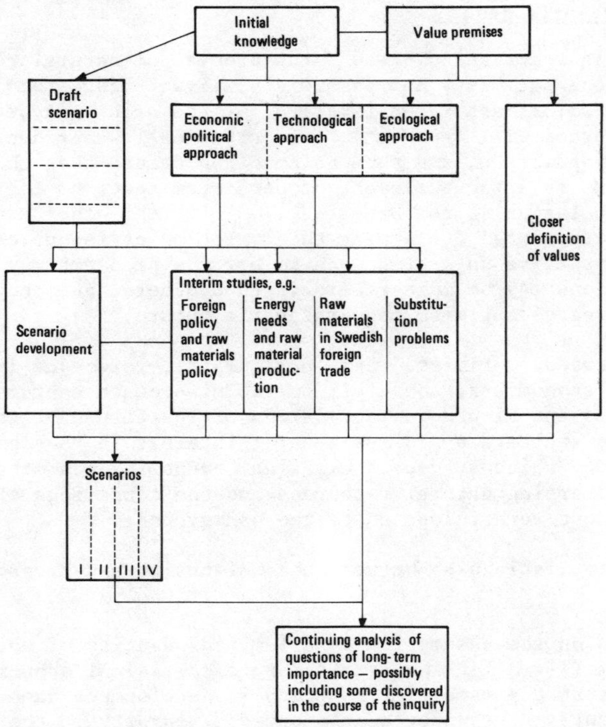

Source: "Resources and Raw Materials - Some Points of Departure for a Futures Study", the Resources and Raw Materials (RoR) Study 1975,(in Swedish)

This did not materialize, partly because, for all our attempts, we found the "method" rather unwieldy. The present report is therefore not primarily based on such "scenarios". (A short description of our work on these scenarios will be found in Chapter 6).

Another question arose out of the desire to communicate our findings to the general public as well as to decision-makers and politicians, namely: what is the difference between a futures study performed in our way and a report from an ordinary ad hoc government commission in Sweden?

There need not exist any strong distinction between the two. This study, like the reports of ad hoc Government Commissions in Sweden, is financed out of public funds. At the same time nothing prevents an official report from focussing on the future. Most official reports however concentrate relatively strictly on a single sector, whereas a futures study takes in a wider field and is particularly interested in link-ups between different sectors of the community. A futures study is also more free to choose its terms of reference, it need not be close to operational significance.

The main purpose of a futures study is not to point to the most probable course of development, i.e. to make a forecast, but rather to indicate alternative futures. In our particular study, the international environment has come to play a more important part than in the majority of Swedish Government official reports. We hope thereby to be able to put Swedish developments into an international perspective, so that the debate concerning our dependence on the surrounding world and the need for international solidarity can be conducted in more concrete terms.

History of a debate

It is not entirely accurate to present the problem of natural resources as a controversy between optimists and pessimists. Even though the 'optimists', confident in the adjustment mechanisms i.a. of the pricing system, and are not afraid that resources will 'run out', they are usually apprehensive of the growth of world population, and they quite often demand that this growth be restrained. There is a tacit element of pessimism here, to the effect that 'if the growth of population is not arrested,...'. On the other hand the warnings issued by the 'pessimists' concerning the depletion of resources may also incorporate constructive suggestions as to how the problems may be solved. Some of these suggestions may be more realistic than others, but they all bear witness to a certain degree of optimism concerning the future.

In the debate between 'optimists' and 'pessimists', reference is frequently made to the classical economists, above all to the nineteenth century English economists Malthus and Ricardo. The visions of the future of the Club of Rome have been said to indicate a growing general interest in the 'grandiose' questions. The economic theory which succeeded classical economics had other preoccupations, and questions concerning natural resources and the conditions of long-term economic development were relegated to the background.(1)

It is customary to distinguish between the Malthusian and Ricardian senses of the term scarcity.

Malthus proceeded on the assumption of a limited quantity of non-renewable natural resources ('land' in the terminology of classical economics) whose quality remained constant at the same time as economic development came to require a progressively greater extraction of resources. Scarcity in the Malthusian sense occurs at the moment when all natural resources have been consumed. Population growth and the speeds of economic growth determine exactly when resources 'run out'. Malthus is thus said to have employed an absolute concept of scarcity.

Ricardo also starts with the assumption of a limited stock of resources, though unlike Malthus he does not assume the quality of natural resources to be constant. Economic development forces man to use resources of progressively inferior qualities. With Ricardo, therefore, resources do not become exhausted all of a sudden, but the cost of extracting them - measured in labour and capital inputs - rises. This relative scarcity concept of Ricardo's does not in principle exclude a successive adjustment to new resource situations.

These two scarcity hypotheses, and least of all Malthus's, are not very well borne out by historical developments. Briefly, technological progress and the possibility of substituting one raw material for another have been the principal counteracting influences.

Nonetheless, at fairly regular intervals in the course of history there have been discussions, based to a greater or lesser extent on these scarcity hypotheses, concerning the exhaustion of resources.

The leading English economist Stanley Jevons raised the question of future coal supplies in "The Coal Question", written in 1865, which aroused a great deal of

(1) It should be observed, however, that the physical shortage of natural resources were not the main problems that interested Malthus or Ricardo. Scarcity of resources was one of many elements in the analysis of economic distribution in society and the conditions attaching to continuing economic growth.

attention at the time. He proclaimed that England's deposits of coal were running out and that the price of coal could be expected to rise very steeply in relation to the prices of other raw materials, the best deposits having already been exhausted. This analysis was based entirely on the Ricardian hypothesis of scarcity. Britain's industrial hegemony, which to a great extent, was founded on coal, was therefore in danger, and Jevons therefore prophesied that the United States was destined to become the world's leading industrial nation. The latter prediction was right enough, but Jevons failed to foresee the development of other types of energy apart from coal and, like Malthus, he made the mistake of basing his predictions on the assumption of a static technology.

The sudden rise in the price of coal during the closing years of the nineteenth century led the Swedish economist David Davidsson to warn against a future coal shortage in an article published in 1900 in the journal Ekonomisk Tidskrift, doubtless influenced by Jevons. Referring to Britain's estimated coal reserves and extraction rate and the general economic decline which it was feared would result from an increasing coal shortage, Davidsson called for a policy in Sweden which would make the country independent 'as regards our fuel requirements'. Needless to say, the predictions made at that time concerning the future coal situation were not particularly accurate.

Another interesting example of advocacy to the point that resources are in danger of becoming exhausted, which, unlike Jevon's, basically was not founded on economic theory, is the Conservation Movement in the United States during the early years of this century whose most important demands were the conservation of natural resources and steps to prevent their exhaustion.

The interests and groupings supporting this conservationist movement cannot be conclusively specified, but it is customarily alleged that they included a large proportion of people within natural science affiliation. Consequently the problem of resources was analysed primarily in physical terms. On the grounds that rising production must mean dwindling resources, fairly cut-and-dried conclusions were drawn concerning the juncture at which one resource or another was going to 'run out'.

But the same movement also included people who employed moral and natural-ethical arguments in support of the conservation principle and who therefore mainly criticized what they considered to be mismanagement on the part of the large scale raw material industries. Indirectly this criticism implied doubt as to whether the market economy gave sufficient consideration to the fact of natural resources being finite. It is not difficult to find similarities between the approach represented by the Conservation Movement and that which in our own times has come to be termed an ecological standpoint.

The movement had a direct political impact. Questioning as it did the possibility of achieving optimum resource utilization through the mechanisms of the economy, it came to advocate a greater measure of planning where natural resource were concerned. Critics accused the movement and its political standpoints of scientific elitism. Advocacy of state controls over natural resources on purely scientific grounds, they argued, was a way of circumventing the normal political process in order to justify state intervention.

Incidentally, President Theodore Roosevelt, who has gone down in history as the President who tried, during the first decade of this century, to strengthen the power of the Presidency and Union vis a vis the big corporations and the individual states, received a certain measure of support from the Conservation Movement.

History also provides examples of optimistic forecasts concerning resources, the best known perhaps being the report of the Paley Commission, presented in 1952. This Commission was appointed by President Truman to investigate long-term supplies of raw materials in the United States. In its report, entitled "Resources for Freedom", the Commission declared that there was no reason to fear a general physical shortage or the exhaustion of resources. Indeed extraction costs were expected to rise, and so were the prices of certain raw materials, but the main problem was 'political availability' rather than physical supply. On the whole, therefore, the Paley report presents an anti-Malthusian perspective.

As should now be clear, a debate on resources can be conducted in various terms and at various levels. The arguments put forward are quite often contradictory. Furthermore, resources and raw materials cannot be discussed in isolation from other great issues concerning the future. It is also clear that values - scientific, political and professional - have a crucial effect on views concerning problems of resources, just as on views concerning other important issues for the future. The values embraced by the authors of this study have naturally influenced both the selection of topics for consideration and the whole argument and presentation which now follows. The concepts employed and the assumptions made concerning relationships are also governed, in the ultimate analysis, by our values.

2 Resources and Raw Materials Defined

Introduction

One cannot discuss any problem without understanding the terms employed in the discussion. This applies particularly to resources and raw materials, because many differences of opinion - for example, concerning the length of time that minerals can be expected to last - are attributable to the use of different concepts.

In the present chapter, the terms are discussed with reference to figure 2. Substances are elements, chemical compounds or mixtures having specific chemical or physical properties. Natural resources are those substances in the soil, water or the atmosphere which in principle can or could be used by man. In the ultimate analysis a natural resource represents the properties possessed by certain substances or combinations of substances, e.g. having a certain electrical conductivity, being hard or being resistant to acids. This becomes particularly apparent when soil, air or water are referred to as natural resources, because the important thing may well be, not the actual substances present in the resource but the properties of those substances - for example, being a basis of biological production (soil, water), or having the capacity to absorb and carry off substances(water).

"Resources" are those parts of natural resources which are known and accessible and which it is believed can, come to be used under foreseeable technological and economic conditions. Reserves are those parts of resources which are already sufficiently known and accessible to be utilized profitably with the available technology. Raw materials, finally, are substances or mixtures of substances which, in one form or another, are used in industrial or other processes in order to manufacture a product.

We may already here observe that the raw material, reserve and resource concepts are not timeless. New technology and changes in the prices of raw materials may have the effect of widening the resource base. The inclusion or non-inclusion of substances among resources depends on the development level, structure or organization of society. In order for something to be a natural resource, both the material foundation and a particular social environment are necessary (the latter in terms of a state of knowledge, organization and needs).

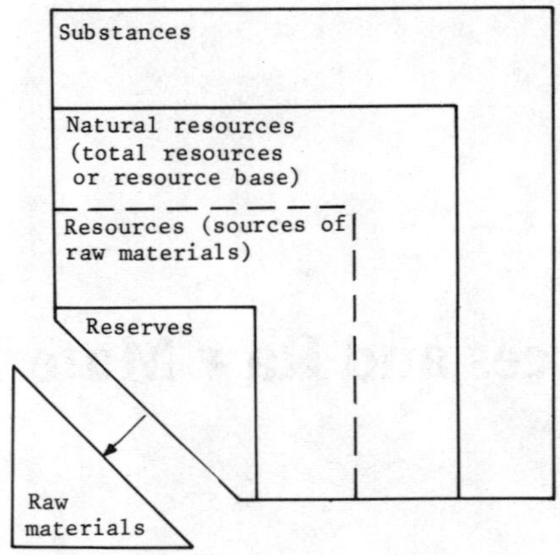

Figure 2. Some central concepts in the debate on resources.

The physical foundations of the resource concept

Resource Base and Civilization

All societies, past and present, rest on material foundations. Gathering and hunting peoples utilized the direct gifts of nature in the form of plants and animals. In this way they obtained food, clothing, shelter from wind and weather and, quite often, the material for different kinds of medicine. The introduction of livestock farming and agriculture meant that more systematic use was made of the same resource base. Instead of looking for plants and animals, many cultures obtained a resource base by controlling, to the best of their ability, the selection of plants and animals which nature provided in their immediate surroundings.

Through co-operation (voluntary or coercive) and by deliberately using his knowledge of the workings of nature, man has been able in the course of history to create various civilizations. The Mesopotamian and Egyptian civilizations were based on intensive irrigation farming (the surplus from which could be applied by the rulers to monumental building projects, for example). The Scythian and, later on, the Mongol empires were partly based on the extensive exploitation of lands which provided grazing for large numbers of horses.

In all of these cases, civilization was made possible by a process of interaction between human knowledge and labour and, on the other hand, the natural base to which that knowledge and labour related. Reference has often been made to the close connection between forms of production and social systems. For example, civilizations based on irrigation ("hydraulic civilizations") require advanced co-ordination of the actions of many farmers and therefore tend to be centrally governed.

So far we have spoken of such human benefits as food, clothing and shelter from wind and weather. The progress of civilization is closely bound up with the development of tools of various kinds, to obtain these and other benefits. It is worthy of note that the "ages" of prehistory are named after the materials of which their tools were fashioned. Even today we have a reiteration of the Stone-Bronze-Iron Age classification in the references sometimes made to the Plastic Age, though of course this will not do as a blanked designation for our time. (There are other alternatives: the Atomic Age, after a certain type of energy resource, the Space Age, after a technology of transport and spatial control, and the Computer Age, after a method of systematically processing information).

Substances

The material basis of society consists of 'substances' and of the energy which in various ways is stored in them. 'Substances' or energy are a resource because in principle they can be used by man for a particular purpose, i.e. they satisfy a human need. Some uses are more direct than others. A certain metal may come to be used because it is a hard, tough substance. This property can be an asset, for example, in an object which is used to split another object in two. Water is needed in various biological organisms, but it can also be used as a lubricant because of its fluidity, as a firefighting substance because of its ability to cool fires and isolate them from oxygen, or as a waste recipient because of its ability to absorb or other wise carry different substances and transfer them from one place to another. Some natural resources, e.g. 'land', are collective designations for many different parts which, taken together, constitute a certain resource. In the case of 'land', resources can stand for 'biological productive capacity' but also for 'area', 'substances present below the surface of the soil', 'soil as a risk absorber' (i.e. the necessity to use large areas of land around for instance nuclear power plants in order to diminish the effects on human society in case of a potential catastrophe).

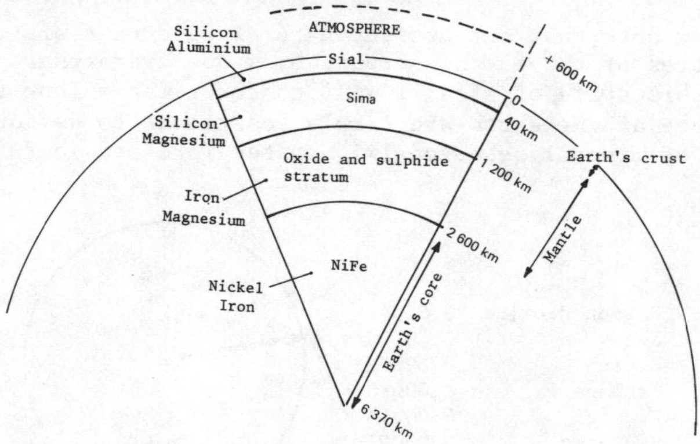

Figure 3. The structure of the globe

In principle the 'substance' of the globe and the energy stored with their help constitute the conceivable resources at our disposal on this planet. They include both organic and inorganic 'substances' in the earth's crust, in the oceans and in the atmosphere. Then there is irradiated solar energy (or other forms of radiation from the universe). Viewed in a more long-term perspective, the substances at our disposal are sometimes taken to include those present on our closest neighbours in space, but this dimension is still considered on a level with science fiction.[1]

The globe is not uniform in structure all the way through. Its outer layer is called the lithosphere and is about 30-60 km thick. The outer stratum of the lithosphere, which makes up most of what we are able to use today, consists of light, granular rocks containing an abundance of silicon and aluminium and is therefore termed SIAL (from the chemical designations of silicon - Si - and aluminium - AI). Further in (down to a depth of about 1 200 km), the proportion

(1) Cf J. Peter Vajk, "The Impact of Space Colonization on World Dynamics", Techological Forecasting and Social Change, Vol. 9 No. 4 1976, p. 361

of magnesium increases. This part of the earth's crust is (somewhat inappropriately) termed SIMA. Closer to the earth's core, the most important substances are iron (90%) and nickel (8%). This is why the earth's core is called NIFE (Ni standing for nickel and Fe for iron). Both temperature and pressure are very high at this point.

It is not only the substances in the earth's crust which, in principle, are accessible. The waters of the oceans and the substances dissolved in them, as well as certain substances present in gaseous form in the atmosphere, also have to be taken into account; see figs. 4-6. Atmospheric nitrogen is already being exploited today in industrial processes to make chemical fertilizer. As far as substances in the oceans are concerned, we are already using common salt, and magnesium (a light metal) and bromide may come to be extracted from the sea in larger quantities than are already being extracted today. There is also talk of other substances such as uranium and potassium being extracted from sea water.

Most substances would be very difficult to extract for human requirements if the geological processes of nature had not concentrated them here and there. Man is able to utilize certain metals because nature has initiated the process of concentration which man then completes by processing ore, smelting it and finally manufacturing material from it - in a steelworks, for example.

Summing up, we may note that our present natural resources are taken from the first few kilometres of the earth's crust, from the hydrosphere and from the atmosphere; and this state of affairs will continue for a long time to come. The mantle and core of the earth are simply too remote to be counted as a potential source of mineral resources within the foreseeable future.

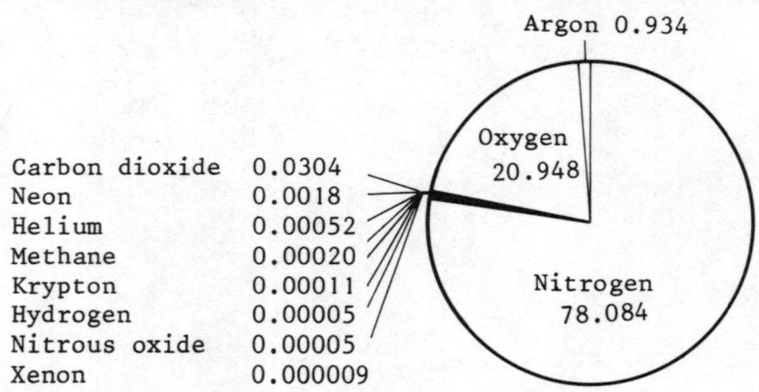

Figure 4. Air (percentage by volume content of dry air)

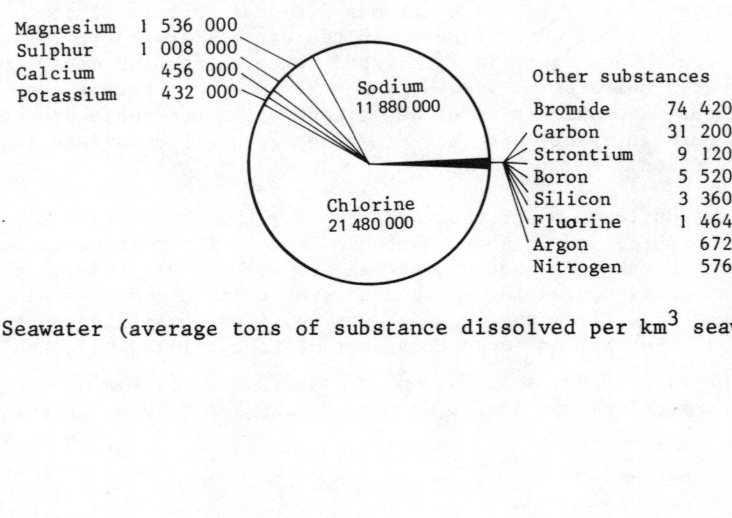

Figure 5. Seawater (average tons of substance dissolved per km^3 seawater)

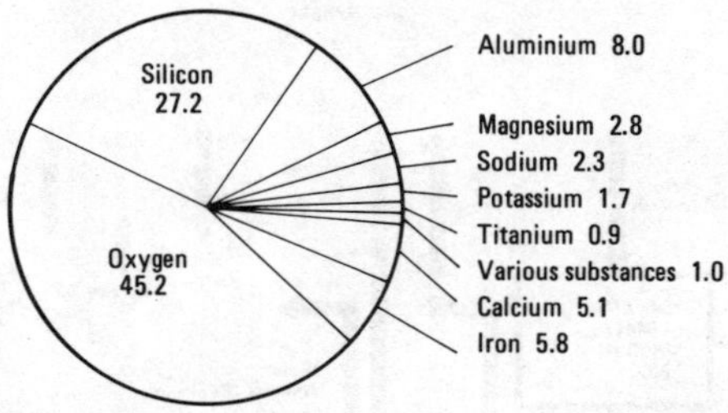

Figure 6. The earth's crust (contents in percentages by weight)

Biological Substances

So far we have mainly been considering inorganic substances. The biosphere, the life-supporting stratum surrounding the earth, also contains organic substances. The substances are made up of inorganic elements and have been built up by the life processes. 'Living matter', i.e. the substances bound up in plants and animals, is known as biomass. The biomass of dry land, i.e. the total weight of the plants and animals now living on the land, is roughly 3×10^{12} (three million million) tons. Animals account for merely one per cent of this total. Trees in various forms make up the greater part of the biomass of plants.(1)

One of the characteristic properties of the biosphere is that the substances taking part in these processes move in cycles of various kinds. The carbon cycle is an important example: see fig. 7. A description of the carbon cycle incorporates time scales ranging from hundreds of millions of years to a mere matter of weeks and months in the rapidly pulsating life of the surface strata of the oceans.

The processes involved can be geological, chemical, physical and biological. Volcanic eruptions and eruptions on the sea bed, for example, can transfer material from the interior of the earth to the surface or to the atmosphere. The continuous weathering and chemical dissolution of the bedrock contributes in its way to the transfer of large quantities of matter to the hydrosphere, either in dissolved form or in the form of deposits on the sea bed or the beds of lakes. The cycles of most substances are powered by solar energy, and sunlight plays a vital part in the building up of biological substances.

The carbon cycle is particularly interesting because it plays such an important part in the biological processes. All life is built up round the unique chemical properties of the carbon atom. In plant photosynthesis, which requires a supply of energy, carbon dioxide is absorbed from the air, reacts with water in a chemical reaction giving carbohydrates and oxygen.

In this way nature builds up more complex molecules (e.g. carbohydrates) from simple components, thus supplying man with a 'commodity' (e.g. 'food') which is of superior quality to its constituent elements.

The contrary process takes place, for example, when animals 'burn' carbohydrates in their body cells, which they have to do in order to live. In this process which is the opposite of photosynthesis, oxygen is absorbed from the air through respiration. The reaction with the carbohydrates produces carbon dioxide and water. Thus the air which you breathe out contains more carbon dioxide than the air you breathe in.

(1) NOU 1972:1 Bruken av Norges naturresurser, ("The use of Norway's natural resources"), Universitetsforlaget 1972, p. 42.

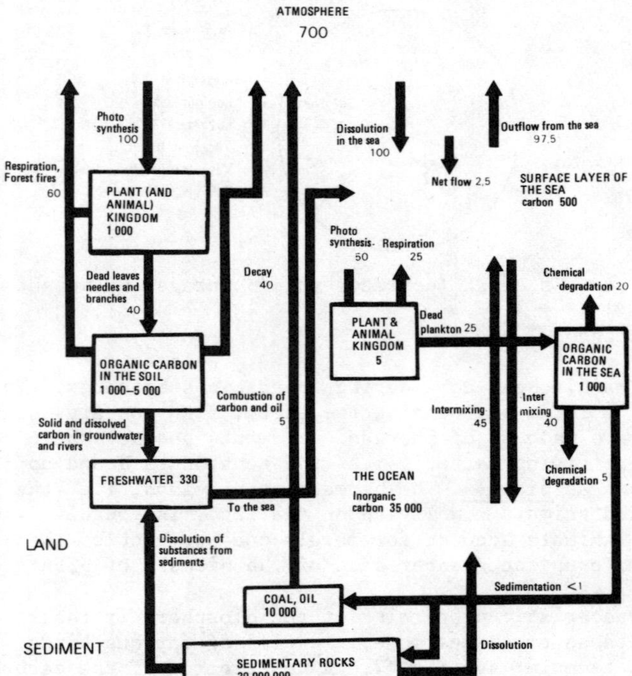

Figure 7. The carbon cycle. The sizes of the reservoirs are given in 1 000 million tons and the flows between reservoirs in 1 000 million tons per year.[(1)]

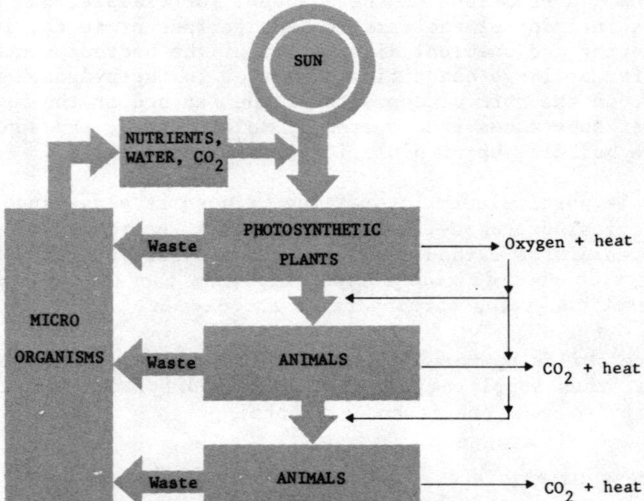

Figure 8. The natural cycles of food and energy. Source: "Kretslopp", 1976. CO_2 is the chemical formula for carbon dioxide.

(1) Figure from Bert Bolin, "The Interaction of the Global Cycles - Balance or Imbalance", published (in Swedish) under the title "Kretslopp" in the 1976/77 yearbook of the Swedish Natural Science Research Council.

The Classification of Natural Resources

It is with these conditions in mind that reference is quite frequently made nowadays to 'renewable' as opposed to 'non-renewable' resources. What this means is that the more sophisticated or 'higher quality' molecules referred to above are 'renewed', at least so long as the cycle remains unbroken. New trees are constantly provided in this way, so that timber, viewed in this light, is 'renewable'. The antithesis implies that 'non-renewable' resources, e.g. minerals, can never be replenished once they have been extracted.

Because this idiom is very widespread it is worth spending some time analysing it in more detail; it can easily give rise to fallacies. It is right and proper that a distinction should be made between natural resources consisting of biologically constructed complex molecules (e.g. carbohydrates) and simple substances (e.g. metals). These might perhaps be termed 'photosynthetic resources' and 'simple resources' respectively. But the degree of 'renewability' is not necessarily a direct expression of this. Mineral oil, for example, is derived from plants and animals of different kinds that lived a very long time ago; in other words, it is based on photosynthesis. But this particular resource is not renewed fast enough to make it 'renewable' for practical present-day purposes. Thus photosynthetically based resources can be 'non-renewable'.

The term 'renewable' can also imply that the resources in question are included in a cyclic chain of surveyable duration. Mineral oil and rape oil are both 'photosynthetic resources', but mineral oil is considered 'non-renewable' because, unlike rape oil, it forms so slowly. In this case, then, the two designations are based on a distinction between the time scales of the cycles.

Sometimes the term 'flow resources' (i.e. resources included in a 'flow') is used as a synonym of 'renewable resources', while 'stock resources' corresponds to 'non-renewable' resources. This terminology is more exact because it refers to a single basis of assessment, namely the time it takes to recover the resource via the cycle.

The quality 'included in a natural biological context' has often been connected to the idea of 'cycles'. But as we have already seen, there are other kinds of cycles, e.g. those of a purely geochemical nature, which can take several hundred million years. Where certain metals are concerned - the iron incorporated in vehicle bodies, for example - a great deal of material is re-used in a 'technically organized cycle'.

The parts which disappear as a result of corrosion and wear are in practice lost because the atoms become so widely scattered. But a dispersion of, say, metallic atoms which today would be considered complete because (for reasons of economics or energy supply) we have no feasible technical for reconcentrating, may possibly be regarded as less than complete at some future date. This may be achieved, for example, by employing a biological link in an otherwise technically organized chain for the recovery of the metal. Experiments have taken place in which bacteria have been used to concentrate uranium. Certain ascidians (marine creatures) have been found capable of concentrating the metal vanadium. Thus new types of concentration processes may eventually cause substances which today are considered hopelessly scattered and therefore classed as 'non-cyclic resources' to be classed as 'cyclic resources' instead. Generally speaking, it is concentration that turns a substance into a resource, so that once a substance has been concentrated (either by nature or as a result of human technology) it should not be dispersed again unnecessarily.[1]

Certain 'non-renewable raw materials' of a more complex nature, e.g. oil, may possibly be produced synthetically in the future by technological alternatives to or accelerations of the natural cycles which have created these 'stored resources'. If this happens, the resources in question will cease to be 'non-renewable'. Without going any further into the question of how likely this is to happen, we may observe that classification can depend on the level of technology in a society.

Another aspect is that 'renewable' resources are closely interlinked with 'non-renewable' ones. For example, large quantities of 'non-renewable' resources, such as mineral oil to run agricultural machines or to produce artifical fertilizer are employed nowadays in the production of plant crops and timber, i.e. to produce 'renewable resources'.

The renewability of renewable resources also depends to a great extent on the possibility or impossibility of preserving the biological basis intact (for instance, by saving it from the build-up of toxic chemicals, erosion, etc). In this sense they can be termed 'conditionally renewable'.

Thus in referring to 'renewable' as distinct from 'non-renewable', you have to be clear in your own mind as to which of the following observations you intend to make:

(1) the substance has originated biologically (photosynthetically based, e.g. wood, mineral oil); or otherwise

(2) the substance forms part of a cycle which, considering the amount of material involved, means that it is 'recreated' within a reasonable space of time (a 'flow resource', e.g. wood but not minerals or mineral oil);

(3) the substance is included in a natural cycle but not in a technical cycle. This involves an assessment of capacity for reconcentrating 'dissipated atoms'.

Cognitive foundations of the resource concept

So far we have confined our attention to the physical basis of natural resources. But resources may be 'resources' solely in relation to a particular society. There must be demand for the resource in society, and society must have the knowledge and capacity to utilize the resource. Unless these conditions are satisfied, the substances in the natural environment will be mere phenomena and not a 'resource'. People who live by gathering food must know which roots and berries are edible and when and where they are to be found. Without this knowledge, the roots and berries are not a 'resource'. The ferrous ores of the Swedish province of Norrbotten, which have a high phosphorus content, did not become a resource until technology (the Thomas process, invented in 1878) made it possible for this type of ore to be used in the manufacture of iron. Coal in medieval England was not the resource which a series of technical improvements were to make it in the seventeenth and eighteenth centuries. Until scientists found out about nuclear reactions, uranium was long regarded as a worthless metal; it became a natural resource - though not a universally accepted one - after this scientific knowledge had been

(1) This is a paraphrase of the second law of thermodynamics, which shows that the 'order' or 'degree of concentration' of a resource is an important quality.

gained and after the appropriate technology had been developed.

But it is not necessarily technical factors that convert natural substances into resources. Lack of demand may be due to cultural factors. In some civilizations people refuse to eat horsemeat. In others, for example in Moslem countries, pork is considered 'unclean'. Between fifty and seventy years ago, to take another example, Swedes generally thought that mushrooms were only fit for cattle to eat. Today mushrooms are grown commercially for human consumption, and they may develop into a really important food product, because certain mushroom species are much more efficient than livestock as a means of converting raw materials containing carbohydrates into protein.

It is quite possible that certain cultural taboos can be traced back to a risk assessment built into the civilization in question. For example, it has been observed that the rejection of pork in certain subtropical countires may be due to the risk of trichina. The rejection of mushrooms as a food resource could be attributed to the fact that certain fungus species are poisonous or harmful, so that experience pointed in favour of avoiding this type of food altogether. The long delay in using waste, e.g. digested sludge, as an artifical fertilizer, thereby turning it into a resource, has doubtless been partly due to a reluctance to take the risk of including pathological micro organisms or certain heavy metals in growing crops without conducting exhaustive experiments first.

The debate concerning the continuation of the nuclear energy programme has shown that large sections of the population in some industrialized countries are prepared to abstain from something which is already a resource in order to avoid risks of different kinds.

There are other factors besides risks which can induce man to refrain from utilizing certain resources. Decisions to refrain from harnessing certain rivers or to refrain from mining operations or the construction of facilities in areas of outstanding natural beauty serve to show that something constituting a resource in the physical, economic and technological sense may be left unused because society attaches greater value to it as a resource <u>in some other respect</u>.

Something which has been a resource at one time or another need not remain so indefinitely. Technical progress and changing needs often have the effect of confining a certain form of exploitation to a certain period of time. Until the early years of this century, sea ice was a resource because it could be used for cooling food. The introduction of the refrigerator put an end to this resource. A few years ago the head of the National Swedish Forest Service was able to report that the oak trunks designated for the construction of men-of-war were now ready for delivery. The Riksdag had earmarked about 25000 acres of land for this purpose in 1829. This rather off-beat example confirms the general rule that a resource does not become a resource until it is related to a certain kind of use in human society. And that use is only defined in a particular social situation which is subject to gradual change.

Resources and Mineral Reserves

We have already quoted a number of historical instances to show the importance of technological and, consequently, economic factors in determining what is available of a natural resource and what is not. This applies particularly to minerals.

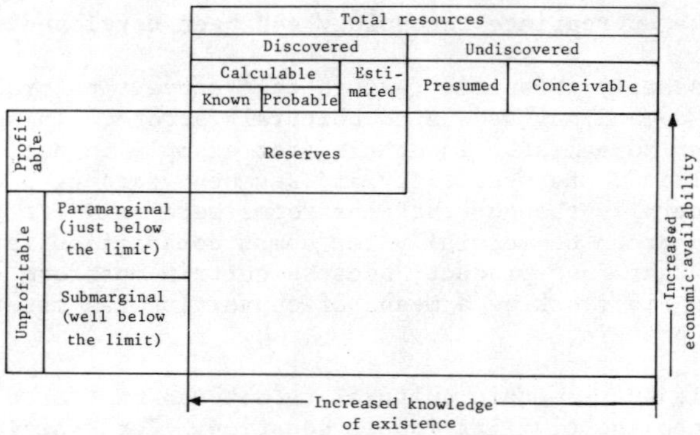

Figure 9. Common classification of minerals and energy assets.(1)

Figure 9 shows assets in a two-dimensional guise. As can be seen, reserves are assets which have been discovered and are worth exploiting at present day prices. If mineral prices go up, reserves will expand further down the chart. Conversely, if extraction costs rise without prices following suit, the reserve panel will contract. Technological improvements can make reserves out of what used to be unprofitable assets. Increased propecting can cause the reserve panel in the chart to expand to the right.

Potential resources are hard to define. Very roughly speaking, a potential resource is a concentration of a substance which is greater than that existing in the surrounding environment and which makes economic exploitation either possible or potentially possible. This is a very wide definition, for the 'potentially possible' will be partly determined by the technology at our disposal in the future. Consequently many estimates of potential resources are really a kind of tacit technological assessment.

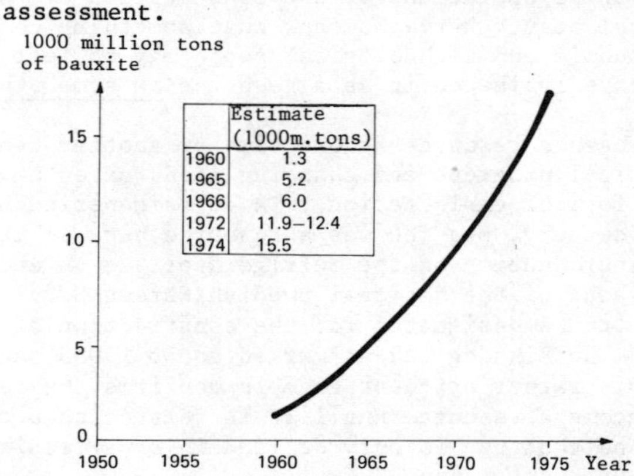

Figure 10. Growth of the world's estimated reserves of bauxite

Source: "Aluminium - A Raw Material Chain", the RoR study, report 305, 1976. (In Swedish)

(1) For a closer discussion of the implications of these concepts, see "Renewable Raw Materials" the RoR study, report 311, 1977. (In Swedish).

PLENTY OF ICE THIS SUMMER! HEAT GIVES COLD DIRECTLY: AN IDEAL COOLING APPARATUS

CENTURIES OLD PROBLEM SOLVED BY
TWO YOUNG STUDENTS AT THE ROYAL INSTITUTE OF
TECHNOLOGY

EPOCH-MAKING DEGREE THESIS

AN INEXPENSIVE, ROBUST AND EASILY OPERATED COOLING
MACHINE FOR HOUSEHOLD USE — ON THE MARKET SOON

In February 1925 the Swedish newspaper Svenska Dagbladet published a report on the new "cooling machine", with an enclosed tubing system and with no moving parts, invented by Baltzar von Platen and Carl Munter, two young graduate engineers. As a result of this invention ordinary sea ice became redundant as a resource for cooling purposes. Until then, shops, restaurants and private homes had received regular visits from big, beefy "icemen" wearing thick leather aprons to keep out the cold and carrying heavy blocks of ice. Natural ice was an important commercial commodity, and it was actually exported from Sweden, mostly to Germany, Britain and the United States. Ice cutting was one of the staple occupations of old people at the Stockholm Workhouse. (Picture: Stockholm City Museum. These two pages are taken from an exhibition catalogue, "Technology and Society", published by the Museum in 1977).

As can be seen from table 1, however, certain common metals occur in relatively heavy average concentrations.

Table 1. Percentage contents of various metals in the earth's crust and in reserves

Metal	Average in earth's crust (1)	Presumed minimum content of viable deposits (2)	Concentration factor(1)
Aluminium	8 %	25 %	3
Iron	5,00 %	25-30 %	5-6
Manganese	0,1 %	35 %	350
Chromium	0,02 %	30 %	1500
Zinc	80 g/t	4 %	500
Copper	70 g/t	0,7 %	100
Lead	16 g/t	4 %	2500
Silver	0,1 g/t	500 g/t	5000
Gold	0,005 g/t	5-10 g/t	1000-2000

(1) The natural concentration required in order for the deposit to be viable as things stand today: column (2) divided by column (1).

Source: The Norwegian resource inquiry, "Norges resurssituasjon i global sammenheng". (NOU 1974:55)

To the above figures must be added the possibility of extracting minerals from seawater, in which there are a very large number of substances dissolved. Magnesium, bromide and salt are already being extracted, and uranium too is being extracted on an experimental basis. In the great majority of cases, the average concentrations in the sea or in the earth's crust do not present realistic alternatives even on a very long-range basis.

The important thing is that some of the commonest and most important minerals are present in relatively heavy average concentrations in those portions of the globe which are accessible to man.

Clearly, therefore, reserves are not static, and they do not necessarily diminish simply because we continue using them. Even though the annual extraction of bauxite (the most important aluminium-bearing mineral today) is constantly rising, estimates of the world's bauxite reserves have developed as shown by figure 10.

But that which is true of the aggregate reserves of certain metals throughout the world does not apply in all cases. Certain individual mines have been closed down because, after a period of extraction; there have not been any more profitable assets available for exploitation. Even at this local level, however, discoveries of new assets have often made up for the exhaustion of old ones. In many cases this has followed technological development work which has made it possible for extraction to continue with undiminished profitability. This is illustrated in figure 11, by an example referring to the mining activities of the Boliden mining company.

Although annual output rose steeply between 1925 and 1975 (the dashed line), constant discoveries of new deposits kept the ratio between ore base (reserves) and annual output roughly constant; in other words, the estimate remaining 'life expectancy of mine' remained unaltered despite extraction. This is partly

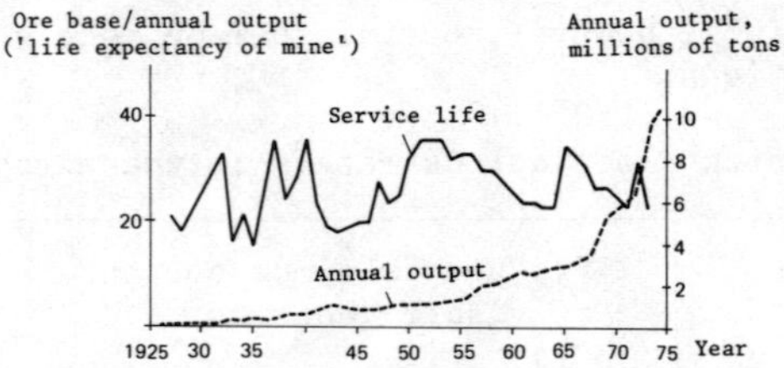

Figure 11. Development of the ratio between ore base and annual output in the Skelleftea minefields.(1)

because the Boliden Company have planned their output capacity in such a way as to attach to each area of ore body where activity is going on a continously updated planning horizon of at least twenty years as a rule.

It is a common fallacy that a declared ore base which is sufficient for twenty years' extraction means that exploitation will be discontinued at the end of the twenty years. This is far from certain. Changing economic conditions may cause a mine to become unprofitable far earlier. On the other hand there have been cases of mines being worked for far longer than was originally anticipated - sometimes for centuries.

It would of course be premature to conclude that continued supplies are automatically assured, especially in the long term. The 'price' paid for maintaining the ratio between ore base and annual production in the case of Boliden, for example, has been the redoubling of efforts to discover new deposits. The graph of the number of known ores in the Skelleftea minefields between 1915 and 1965 has, quite simply, levelled out; see figure 12.

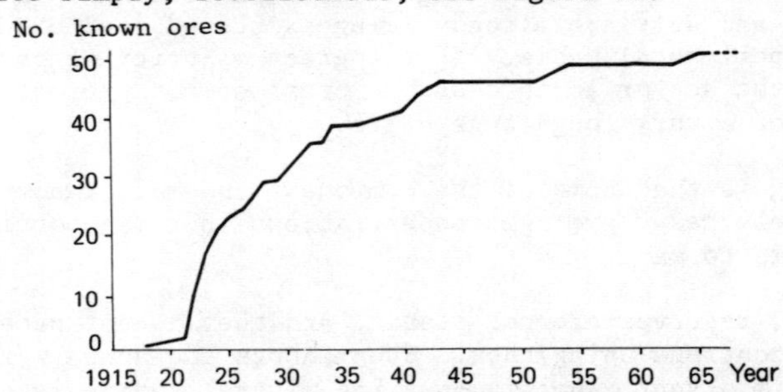

Figure 12. Numbers of known ores in the Skelleftea minefields, 1915-1965.

Source: Sigvard Ando, "Thoughts on the Future of Swedish Mining", Svenska gruvforeningens Meddelande no 140, 1975. (In Swedish).

The target of a constant ratio between ore base and annual extraction has not been achieved everywhere in Sweden. For example, there are places in central Sweden where the ore base has contracted with the passing of years to give a life expectancy of the mine of about ten years. Many mines have been closed down altogether.

(1) The Skelleftea minefields is a common name for an area in the north of Sweden where mining activity is going on in the same type of geological formation. The mineral is of a sulphide type holding copper, zinc, lead and to a lesser extent gold, silver, cadmium and mercury. Economically copper is the most important metal and it is found varying concentrations from below 1% up till around 2%.

Raw materials

Putting it very simply, raw materials, as we have already observed, are substances or mixtures of substances used in one form or another in industrial or other processes to manufacture a product. In other words, raw materials are natural resources utilized in the economic system.

Normally the term raw materials refers to one of the first stages of the upgrading process between resource and end product. Figure 13 shows the processing chain from the unextracted mineral bauxite to finished aluminium sheet. Bauxite can here be regarded as the raw materials for the entire chain, but it is possible, for example, to refer to aluminium oxide as the raw material for the electrolytic production of primary aluminium. The finished sheet can itself be described as raw material for the manufacture of a saucepan or for cladding sheet. In this book we shall use the term raw materials primarily with reference to products of the primary sectors, viz agriculture, fisheries, forestry and extractive occupations such as mining. Sometimes, however, this limitation breaks down. For example, Sweden imports very little bauxite but all the more aluminium oxide and crude aluminium, and it would therefore be a mistake to regard bauxite imports as a reflection of Sweden's actual total demand for raw aluminium.

Thus a hard and fact division into raw materials, semi-manufactured goods and finished products is not possible when goods at different processing stages are exported and imported. The 'economic cycle' of raw materials makes it impossible to use the same single, static definition in every context. The concept of raw materials must therefore be tailored to the question which is being investigated.

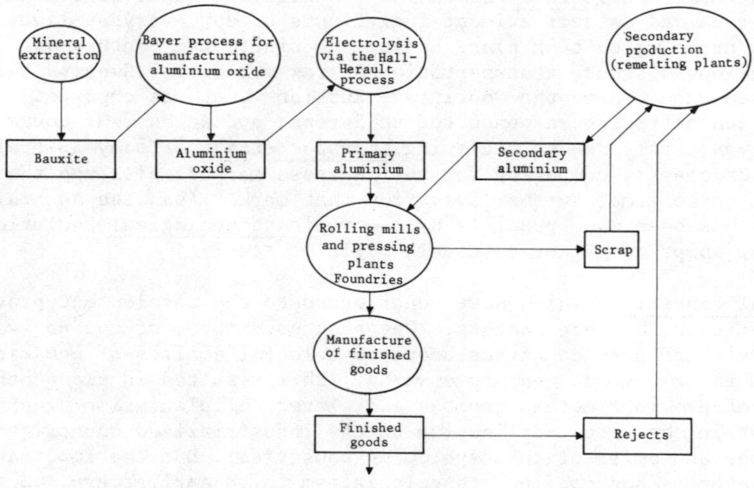

Figure 13. Aluminium production - from bauxite to finished metal
Source: "Aluminium - A Raw Material Chain", the RoR study, report 305, 1976. (In Swedish)

The Declining Share of the Primary Sectors of the Economy in Employment and Production

During earlier periods of history, a large proportion of the inhabitants of the industrialized countries were employed in sectors of the economy which were directly concerned with raw materials. In 1870, for example, 70 per cent of the population of Sweden was employed in agriculture. During the reign of Gustav

Vasa (1523-1560), the Falun copper mine played a tremendously important part in the national economy, to quote another example. At local level there was nearly always a very close link between economic activity and some form of raw material base. The bruk, a distinctively Swedish industrial agglomerate founded on supplies or ore, timber(for fuel) , reducing agents and hydro power, provides an important example of this close connection between society and raw material base in an earlier historical period.

Swedish farming today produces a good deal more than it was doing in 1870, and yet it employs only a small percentage of the national population. This is a global tendency. The proportion of the working population employed in agriculture is customarily used as a guage of a country's degree of development or at least of its industrialization. Thus the majority of people in the industrialized countries are no longer personally employed in activities directly relating to raw materials.

During the twentieth century, we have gradually grown accustomed to the idea and the fact that industrial activities can be based practically anywhere, irrespective of the location of resources. Nowadays Arabian oil is refined e.g. in the small Swedish town of Stenungsund and South Korean tungsten is processed in Sandviken (renowned for the products of high quality steel produced there). This has had the effect of obscuring the raw material base of society; raw materials have been sublimated in the general consciousness because the majority of the raw materials used in our everyday lives are no longer produced in our immediate proximity.

The following presentation gives some of the many factors which have made this possible. International trade has caused raw materials that were once beyond our reach to be regarded as self-evident ingredients of our everyday lives. The great expansion of trade which took place during the nineteenth century was primarily based on the long-distance transportation of raw materials; Swedish timber, pulp and paper were exported to the Continent; Russian wheat was exported to Western Europe; Chilean saltpetre was exported to Europe, and so on. Of course, this development was partly due to declining transport costs. Today Australian and Brazilian iron ore are competing for the Japanese market, although the Brazilian ore is about three times further away from that market than the Australian product. This in turn has been made possible by the rapid technological evolution from small sailing ships to gigantic vessels built of steel.

The political conditions which have contributed to the development process are closely bound up with these factors. One such condition, of course, was that many of today's affluent countries acquired colonial empires of one kind or another from the sixteenth century onwards. This resulted in among other things, heavy flows of precious metals, spices and, later on, plantation crops which played a part in the industrialization of the industrialized countries and in the shaping of the economies of the developing countries. But the importance of the political factor is not confined to colonialism in an earlier era. International trade as we know it from the post-war era is not a spontaneous and natural phenomenon but the result of a deliberate effort - primarily on the part of those in a position of economic strength - to promote an international division of labour, thereby increasing the flow of raw materials and finished goods across the frontiers.

Another reason why the question of raw materials has faded into the background is the ongoing process of industrialization, which has caused people to consume increasingly processed goods. This applies to food, for example. In Sweden today, potatoes, eggs and fresh fish are practically the only food products to be consumed in essentially the state in which they emerge from the basic sectors of the economy. Meat invariably passes through slaughterhouses, and broilers are prepared, packed and frozen before being dispatched to the wholesalers. Even greater value-added is encountered when we eat at a restaurant or a hamburger stand, for example. There are many stages involved in the upgrading process between ore and finished product - e.g. a refrigerator - and usually the cost of raw materials accounts for only a fraction of the price paid by the consumer.

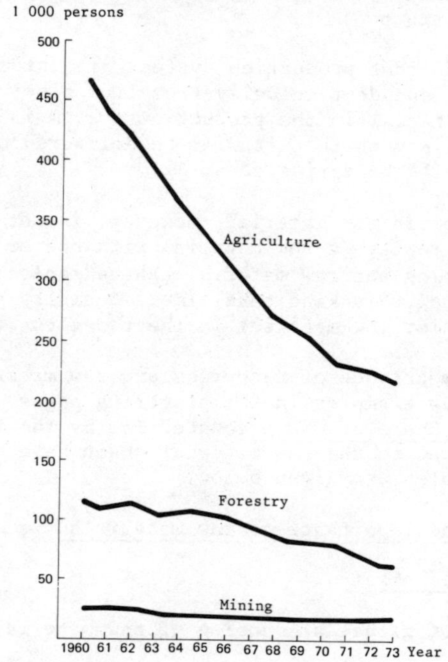

Figure 14. Employment in agriculture, forestry, mining and mineral extraction in Sweden, 1960-73.

Source: Statistiska tabeller no 1974:48, SCB (the Swedish Central Bureau of Statistics).

Consequently, at first glance, raw materials do not appear to have a very important bearing on the standard of living. In 1974 the base sectors accounted for barely 6 per cent of the total value of output in Sweden.

Does this mean that raw materials today have lost the great importance they once possessed? One's intuition says that it must be a mistake to conclude that raw materials are unimportant. A thorough demonstration of their importance is not obtained until prices rise rapidly and/or a physical shortage occurs.

The rising price of oil, which deeply has effected the world economy is a case in point. An example of physical shortage is provided by the natural gas shortage which struck the United States in January and February 1977, with the result that millions of factory workers were laid off, schools were closed and so on. It has been estimated that the cold weather and the gas shortage between them would have reduced the American GNP for the whole of 1977 by 2 per cent or 35000 million dollars, if growth during the remainder of the year had not been so swift as to make up for these losses. But there have also been physical shortages of other raw materials, both during armed conflicts and in connection with raw material crises such as occurred in 1973 and 1974. At one point certain base plastics were quite unobtainable, and parts of Sweden's chemical industry were therefore threatened for a time. In situations of this kind the elimination of a single raw material has repercussions affecting a whole chain of industries.

It has been estimated that mineral production in the United States accounts for 2.9 per cent of GNP. Obviously this figure does not reflect the indirect importance to the US economy as a whole an economy which would collapse totally if mineral production stopped.

This is due to the fact that production systems are interlinked. Production in one sector of industry is dependent on deliveries from other industries. The total importance of a raw material in the production system is not properly brought out until there is uncertainty whether it will be delivered at substantially increased prices or whether it will be delivered at all.

Dependance on any specific raw material, however, is not unequivocal. For example, it may be possible to replace a raw material with one or more others. Or perhaps the product incorporating the raw material can be replaced with a completely new product. But changes of this kind take time. Business enterprises are more dependent on a particular raw material in the short run than in the long run.

How then should the importance of resources and raw materials be assessed? We have already seen a few examples in the preceding pages. Employment in the raw materials sector, the share of GNP accounted for by the raw materials sectors or the value added attached to the raw material chain have been yardsticks of this kind. Some more examples are given below.

Measuring the importance of raw materials

Raw Materials - Volume/Weight

Here is a brief list of global production of raw materials by weight.

Over 1000 million tons. There are only five raw materials whose annual global output exceeds this figure. Sand and gravel output is estimated at about 12000 million tons, petroleum at about 2800 million tons, coal (including lignite) at 3000 million tons and firewood at about 1750 million tons.

Between 100 and 1000 million tons. This category includes a number of well-known staple products - fodder grain (650m. tons approx.), milk (425m. tons), wheat and rice (about 350m. tons each), salt (150m. tons) and meat 115m. tons). Thus total grain production, at about 1400m. tons, qualifies for the top class. But the division of this group into fodder grain and bread grain is justifiable, because the two types are put to completely different uses and yield completely different returns as regards food for human consumption. Iron is the only metal in this class. At almost 500 million tons it is well ahead of the next metal in terms of weight, which is aluminium.

Between 10 and 100 million tons. This category includes a number of foodstuffs but, surprisingly enough, only a few minerals. Bauxite output (for aluminium production) amounts to 75 million tons, while sugar manufacturing corresponds to about 78 million. Phosphate extraction and fishing total about 70 million tons each. Output of vegetable oils and fats is about 50 million tons, and so is output of citrus fruits.

Sulphur output is over 30 million tons and potash output over 20 million. Cotton output totalled around 12 million tons in 1975.

Between 1 and 10 million tons. Among other commodities this category includes such beverages as coffee (5m.tons), cocoa (1.5m.tons) and tea (1.2m. tons). It also includes natural rubber (3.3m. tons), jute and kenaf[(1)] (3.8m. tons), together with copper, lead, zinc[(2)] and asbestos.

If we confine our attention to weight, a number of properties emerge which are neighter immediately obvious nor commonly known. Firstly, energy raw materials are strikingly large compared with all other raw materials. Coal output (lignite included) is six times as great as the world's production of iron, for example. Secondly, 'renewable' raw materials - timber goods, foodstuffs - carry a great deal of weight, literally. Grain output, for example, is two-and-a-half times as great as iron output. Milk production is of the same order of magnitude as iron production. Sugar and bauxite are produced on the same scale. Thirdly, it is interesting to note how far down metals come in the weight league. Iron and aluminium, the two metals occurring most copiously in the earth's crust, are the only ones whose annual extraction exceeds 10 million tons.

Statistical uncertainty apart, however, these figures are open to a number of objections.

One important objection is that we have enumerated raw materials which are in fact used in the production of other goods also included in our list. For example, fodder grain - used for fattening livestock for slaughter - and meat have both been included. This objection can only be overcome by a net estimate of all raw materials - taking into account the finished goods, such as tractors, which are used in the production of these raw materials - so as to arrive at a global, original raw material intake. But this procedure would involve a variety of statistical problems. Many cows are reared on fodder grain. On the other hand the statistics do not tell us the amount of grass consumed by beef cattle.

Raw Materials - Value

We may note before going any further that the impression gained varies according to the denomination in which a resource or a raw material is described. For example, the weight of tin extracted in the world is fifteen times less than the corresponding amount of lead. But tin costs ten times as much as lead, so that in terms of value the two metals are on a level with each other. The extraction volumes of iron and copper are almost 500m. tons and 7.5m. tons respectively (1973). But in terms of the pure metal derived, iron production is worth 33000 million dollars as against 13000 million for copper. Thus the weight of the metal extracted is not always a good criterion of the importance of a raw material.

Figure 15 gives a very rough classification of different groups of raw materials according to value. It goes without saying that the numbers given are very

(1) Kenaf is a fibre material extracted from the east indian hibiscus plant. The fibres are e.g. used for cordage. This type of fibre is sometimes given together with jute fibres in international statistics.

(2) These weights refer to metallic content. Usually it is the minerals containing the metals that are transported. The corresponding figures for mineral weight would be in the 10-100 million ton class.

uncertain. For example, the value of food output has been calculated on the basis of world prices. Even allowing for this element of uncertainty, the immense superiority of total agricultural output over the most important metals in monetary terms is clear. The gap between the two would be wider still if "agricultural output" were made to include other renewable raw materials such as timber products.

Despite the sources of error it involves, the growth of the relative value of output of different goods can provide important information. In addition to straightforward particulars concerning magnitude, it also brings out shifts in the balance between different commodity groups, for example. Figure 16 shows how world output of mineral fuels, metals and industrial minerals has developed since 1900. Mineral fuels (excluding reactor fuels) account for roughly half the value of world output of mineral fuels. (By 'mineral fuels' we mean anthracitic coals, bituminous coals, lignitic coals, oil and gas.)

The other half of the value of the world's mineral production is shared more or less equally by metals and industrial minerals. As is seen, these two categories have changed places during the present century.

It is not altogether easy to assess the value of a raw material in monetary terms. A great many raw materials - especially foodstuffs - never come onto the world market, and the prices applying in different countries are not easy to compare. For most metals the prizes officially given are based only on a limited fraction of all transactions in the world. It is e.g. difficult to get a clear view of the prizes which are used selling and buying raw materials at different stages of processing inside companies. In addition it is difficult to get an overview of the various discounts used in relation to a certain common prize set by a few leading companies.

Then again, different goods occur at different stages of the production chain - milk and cheese, for example. In many cases transport costs have an important bearing on prices, and we therefore need to know whether or not they have been included in figures;the available statistics do not always make this clear.

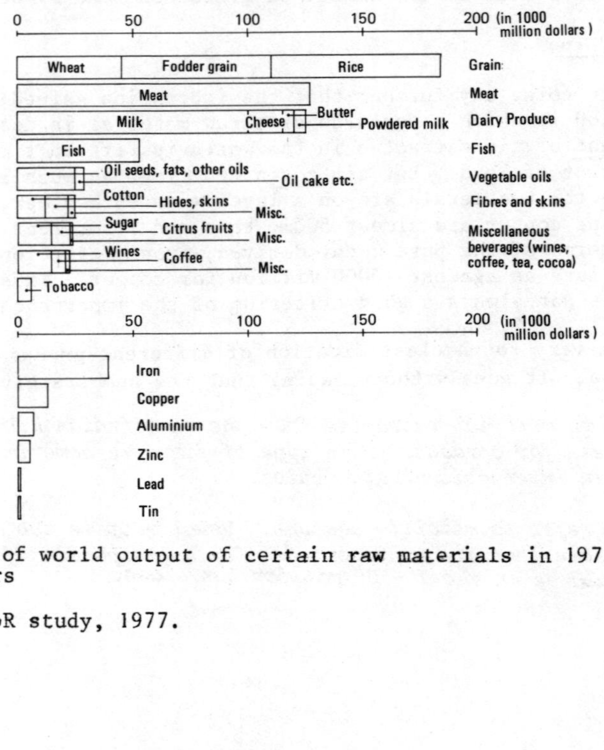

Figure 15. Value of world output of certain raw materials in 1973 in 1000 million dollars

Source: The RoR study, 1977.

What do we mean by 'important' raw materials? What can we most easily manage without, salt or gold?

The top picture comes from the Chinese port of Huiyang in the province of Kwangtung, where salt is extracted all the year round. (Eastfoto via UPI)

Pictured on the right is Mrs Mary Brooks, head of the US mint, surrounded by gold ingots at Fort Knox. (Pressens Bild)

In a futures study, moreover, there is special reason to point out that the market pricing of a particular raw material does not necessarily reflect the long-term supply position or the true extraction costs. Oil prices are a case in point. Current estimates of the available reserves indicate that present oil prices cannot call forward the necessary measures of economization in time to prevent difficulties occurring within the next ten or twenty years. Similarly, the current prices of forest products or silver and tungsten, for example, may be too low to support a reasonable long-term level of consumption.

Classification by Needs

A more universal objection to gauging the social importance of raw materials by their market value as calculated here is that the prices fetched by different raw materials do not necessarily indicate how 'important' they are to the community. A ton of salt can be purchased fairly cheaply, and the value of world output is fairly low. Yet salt is quite indispensable to the living processes of man and beast alike.

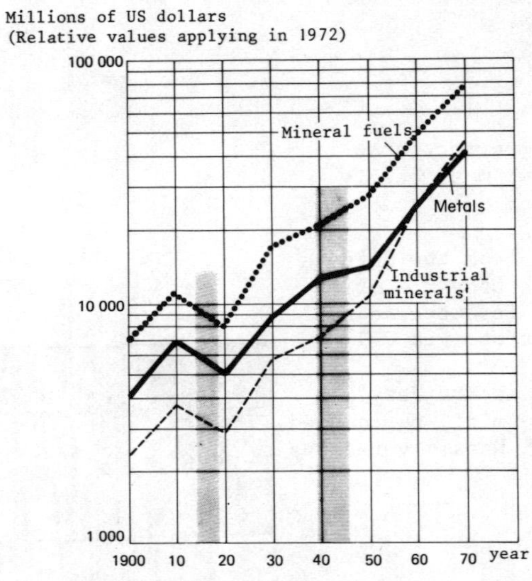

Figure 16. Growth of world mineral production in the 20th century. (Fixed prices)

Source: P.G. Kihstedt, "Raw Materials during an Energy Shortage", special report compiled for the Secretariat for Future Studies, 1977. (In Swedish).

Phosphate extraction (i.e. extraction of minerals rich in phosphates) amounted in the early 1970's to approximately 70 million tons a year, which is about twice Sweden's production of iron ore. This is an impressive figure, but it still does not do justice to the importance of phosphorus as an artificail fertilizer in the production of food. In fact phosphorus is absolutely essential. Substances could be grouped according to their type of use and according to whether they are dispensable or not. This involves several problems, however. Firstly, a number of substances are vitally necessary in certain minor quantities - as tracers, for example. As regards the greater portion of their uses, however, they can be replaced with something else, although this may possibly mean having to accept some functional impairment of the product or an increase in its price. Secondly, it is no easy matter to decide which needs are important over and above the most fundamental ones such as staple foods. Thirdly, raw materials are so intricately bound up with all the various end products that today, at any rate, it is very hard to keep statistical track of all the raw material flows. This situation could be simplified if more efforts were made towards resource accounting.

Some raw materials are important because they indirectly invest other raw materials with properties which the latter would not otherwise possess; for example, the addition of limited amounts of chromium to iron gives a corrosion-resistant metal alloy. Or again, certain substances may be needed as catalysts in certain production processes. In many cases the catalyst is re-circulated into the process after use by technical means and is thus not consumed. However, its presence is vital to the process. In other cases the substance used is not possible to recycle as it is "consumed". Coke is an example of such a substance in its role as reducing agent in the manufacture or iron. Energy raw materials are presumably the most necessary of all these virtually essential 'subsidiary raw materials' for industrial processes. Most materials are interchangeable, but almost invariably energy has to be supplied in order for a process to take place at all.

Thus an assessment of the importance of a particular raw material must show the need which the raw material is taken to meet (e.g. 'pillars which have to support a certain load'), and also the terms in which the critical raw material has been measured (dollars quantity, consumption of energy in production). In figure 17, three materials have been placed in rank order according to total energy consumed in their production and with reference to the task of supporting a load of 1000 tons.

Reference to a difference task might change the rank order in figure 17. There are some cases where raw materials cannot even be compared with each other, Because substitution between them is impossible; for example, you cannot very well construct an aircraft wing out of concrete.

In most cases, therefore, it is impossible to arrive at an unequivocal assessment of the importance of a raw material either in direct quantitative terms (e.g. kilowatt hours for energy) or in monetary terms. This is in particular true as regards natural resources which are prerequisites rather than inputs, e.g. the availability of land or water for agricultural purposes or even the solar energy irradiated over the land in the course of a year. The general rule is that the raw material or the resource must be expressed in a denomination pertinent to the question which is being studied.

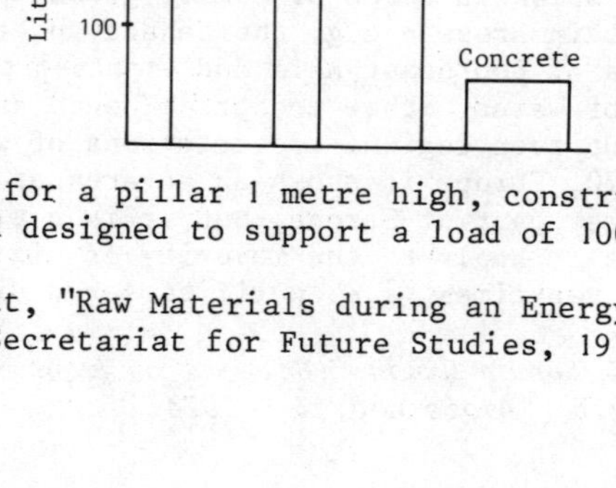

Figure 17. Energy input for a pillar 1 metre high, constructed of various materials and designed to support a load of 1000 tons.

Source: P.G. Kihlstedt, "Raw Materials during an Energy Shortage", special report compiled for the Secretariat for Future Studies, 1977. (In Swedish).

Land area can be measured in square kilometres. If we are concerned with the agricultural value of land, however, the percentage of 'cultivable land' may be more interesting. Figure 18 is one possible breakdown of the earth's resources of land.

Where water demand is concerned, interest may focus on the amounts of water circulating in different parts of the large cycles, because water is a resource which is simply used, not 'consumed'. (The total quantity of water is constant).

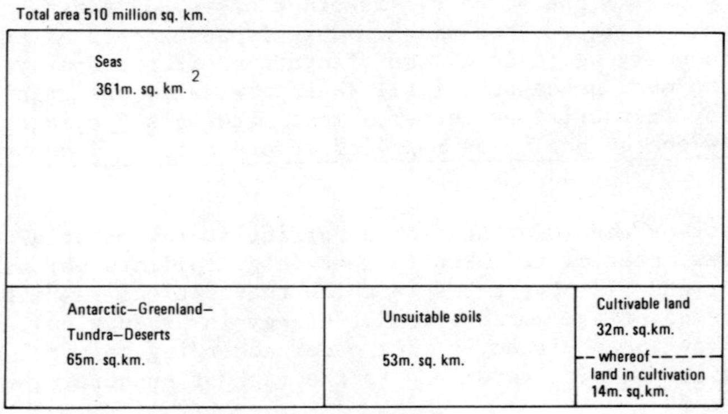

Figure 18. Breakdown of the earth's surface.

Source: UN Document E/CONF. 60/SYM. 111/13.

Figure 19 shows one way of describing a resource of this kind. In this particular case the figures do not convey anything until they have been narrowed down to the amount which mankind can obtain and use from the global cycle. Swedish scientists have estimated, for example, that by the year 2015, mankind will need to use about 20-25 per cent of the world's water supply.(1) Thus we are not interested here in the total amount of water but in the proportion which is available for human use, subject to various standards of purity.

Distribution - an Important Basis of Assessment

All of the above examples concerning ways of describing the importance of a resource or a raw material have been on a global level. But there are many questions where the main concern is with the distribution of a resource and not with its total quantity.
Europe and Asia are far more densely populated than other regions, and they have less reserves of cultivable land. In fact Europe on the whole is considered to have brought virtually every acre of its cultivable land under the plough (Table 2)

It is also difficult to speak in terms of common global problems where water is concerned. Whereas certain areas - e.g. the Sahara and the countries to the south of it, large areas of Southeast Asia and certain parts of South America suffer from a shortage of water, other countries, such as Sweden, are still well off in this respect. But even regional descriptions of water supply can be misleading. In figure 20, Europe is shown as an area of water surplus. This holds good for the greater part of Europe, but it is not true of certain countries (e.g. Hungary), nor does it apply to the majority of big cities. The water of the Seine is utilized seven times (i.e. purified seven times) before it reaches

(1) Malin Falkenmark and Gunnar Lindh: How can we cope with the water resources situation by the year 2015 Ambio nos. 3-4 1974.

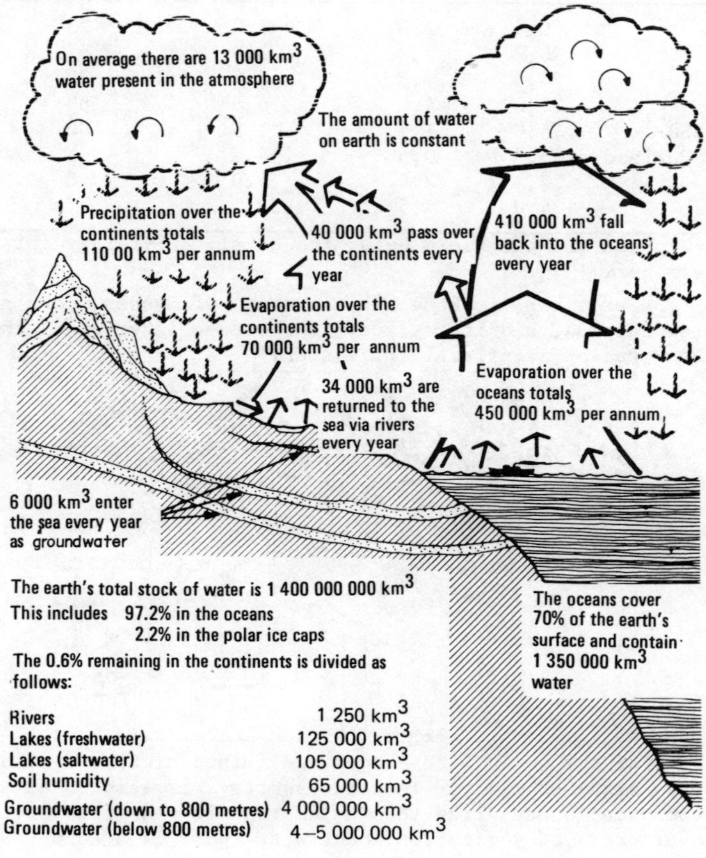

Fig 19 Global Water Cycles

Source: 'Norges resurssituasjon i global sammenheng' (Norway's resource situation in a global context), Norges offentliga utredningar, NOU 1974:55 (Universitets-forlaget) 5, p.108

Water Supply. Evaporation from the oceans totals 450,000 km^3 annually. Of this amount, 410,000 km^3 are returned to the oceans in the form of rain and snow. The remaining 40,000 km^3 are spread over the continents together with the 70,000 km^3 which have evaporated from dry land. Of the 110,000 km^3 precipitation over the continents 34,000 km^3 of water is brought to the oceans by rivers and 6,000 km^3 by groundwater streams. Long-term water supply is determined by run-off and not by the amount of water in the seas, in the air or below ground.

Table 2. Cultivated/cultivable acreage compared with population density.

	Area, m.km²	Cultivated area	Cultivable area	% cultivable area in cultivation	Population density per km²
Africa	30,5	2,0	8,0	25	12
North America	20,0	2,0	3,5	57	11
Latin America	20,5	1,5	6,5	23	15
Asia	27,0	4,5	6,5	69	78
Europe,excl.USSR	5,0	1,5	1,5	100	95
Australia,New Zealand	8,0	0,5	2,0	25	2
USSR	22,5	2,5	5,0	50	11
World	134,0	14,5	33,0		28

Source: Darin-Drabkin and Lichfield,Goldsmith et al: A Strategy for Resources, UN Statistical Year book 1974.

the Atlantic. The province of Skane, in the south of Sweden, does not have a noticeable water surplus, and it may incur a shortage later on if there is an increase in the amount of artificial irrigation.

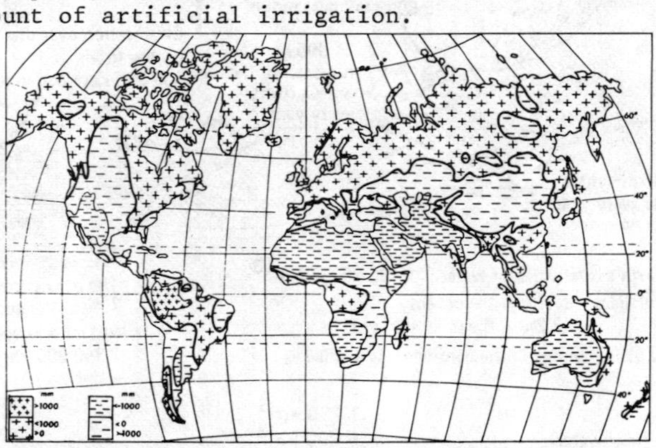

Figure 20. Areas of the world with a shortage (minus sign) and with a surplus (plus sign) of water. (A water shortage is said to exist if precipitation supplies less water than is required for the proper maintenance of vegetation; otherwise there is a surplus).

Source: M. Falkenmark, G. Lindh: 'Water Resources and Society in Different Regions', the Committee for Futureoriented Research, 1976. (In Swedish).

International Dependence

Figure 21 illustrates the problem of distribution between industrialized and developing countries, with reference to the production and consumption of natural and synthetic rubber. Natural rubber as a raw material fits the common pattern of a great deal being produced in the developing countries while most of the consumption is in the industrialized countries. Table 3, Appendix 4, (p. 186) contains particulars concerning annual output and consumption of some minerals.

The distribution of resources is closely bound up with a given country's dependence on foreign trade in particular commodities. Malaysia depends on natural rubber for over a quarter of its national revenue. Between 1972 and 1974 Chad derived almost 70 per cent of its export earnings from cotton, and the corresponding figure for the Sudan was 57 per cent. Lesotho, similarly, depends on exports of wool for 50 per cent of its export earnings, and Bangladesh is almost equally dependent on exports of jute. Burundi derived 83 per cent of its revenue from coffee exports, the corresponding figures for Uganda and Colombia being 67 and

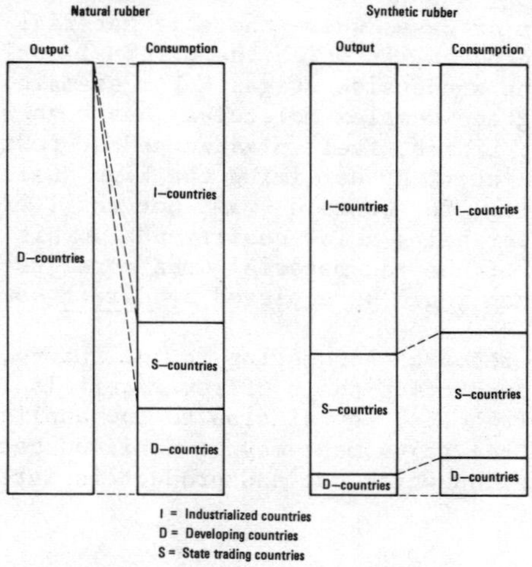

Figure 21. Percentages of world output and consumption of natural and synthetic rubber in 1975.

and 52 percent respectively. Thus many of these countries depend on a single raw material for more than half their export revenues. Even if the raw material in question (e.g. jute or coffee) is not vitally necessary at the global level, it is extremely important in the short run and to the exporting country. Thus when assessing the importance of a raw material, we must indicate what it is important for, who it is important to and, sometimes, for how long a period of time it is important.

It follows that the importance of a raw material cannot be adequately described in mere quantitative terms such as dollars or kilos. The use made of the raw material (or the indispensability of the raw material for a particular purpose) does as much to determine its importance as different production conditions (e.g. the 'geographical accessibility' of metallic ore, the energy expenditure entailed in production or transport and so on). Moreover, there is no reason to suppose that the price of strategically important raw materials, i.e. those which are liable to run short eventually and which are very difficult to replace, is a true reflection of their long-term availability.

Qualitative Aspects and the Importance of Raw Materials

A single raw material can have many different functions, depending on its quality. The following two examples will serve to illustrate this problem.

In Sweden water of the highest quality is used in many places for drinking, for agricultural watering and for cleaning purposes. Other countries often have double water mains, with one main supplying drinking water while the other carries water of poorer quality. Secondly, it has been observed that natural gravel ought in future to be reserved for sophisticated purposes (above all for concrete ballast) because supplies are dwindling.

In some respects, there is a parallel here with the viewpoints that have been expressed in the debate on the need to avoid unnecessary use of 'high quality energy' (e.g. electricity) for low quality purposes (e.g. residential heating). Raw material quality is less easily defined than energy quality. But whatever the definition, the long-term principle must be to avoid unnecessary wastage of high quality raw materials.

There are quite a number of cases where the same material could conceivably be applied to a chain of purposes, in which the original qualities of the raw material are utilized in different successive stages. For example, the raw fibre derived from timber (i.e. high grade complex molecules) has high quality properties which can be put to many uses: fibres, fuel, plastic and so forth. The appropriate procedure may then be to start by utilizing the high quality aspect of the material - in paper making, for example - and not until afterwards to utilize its fuel potential, the latter being a low quality use. This sequential utilization of the original quality of the raw material does more justice to the potential of its complex molecules than would be achieved by <u>direct</u> combustion.

With a view to possible resource accounting in the future, consideration should be given to the entire production chain of raw materials (indirect resource consumption and transport costs) in relation to the applications of raw materials. In this way a better mutual adjustment may be achieved between the raw material, its quality and the function which the end product is intended to perform in a long-term perspective.

3 Global Demand

Introduction

The preceding chapter was devoted to a general description of the volume of raw materials produced and consumed in the world today. How much raw material will be needed during the next twenty-five years?

Clearly, the answer to this question depends on what we mean by need. Things which persons in rich countries like Sweden regard as necessities today - many years of education, a high standard of medical services, a 'reasonable' level of consumption - are well above the standard at which only our fundamental needs - food, clothing, housing and employment - are catered for.

There are today at least 500 million people in the developing countries suffering from severe undernourishment. About 40 per cent of the inhabitants of the developing countries are living in what is termed absolute poverty.

It is arguable that the only needs worth considering are those which are truly fundamental. The strategy adopted by the World Employment Conference in 1976 to provide for 'basic needs' is in fact aimed at focussing all efforts on the eradication of absolute poverty. This means catering for needs which can be said to form part of a 'reasonable' minimum standard and which can be specified, for example, in terms of necessary calory intake per person per day. It is, however, a difficult undertaking - but an important one - to try to identify truly fundamental needs. Needs of these kinds are on different levels in different countries according to their state of development, cultural environment,etc. To the suburban citizen of an industrialized country for example, a car may be considered, and in many cases may actually be, a necessary article because the entire community has been designed for individual transport.

People's subjective views concerning a 'reasonable' standard of living are usually well above a basic level of this kind, at least in the industrialized countries. The expression 'necessities of life' as interpreted in Sweden can, for example, be taken to include considerable quantities of food, cigarettes and amusements like television - things which cannot be considered necessary in any strict sense of the term. Thus there is a difference between basic needs and needs connected with the possibilities of participating in the life of the community on fairly reasonable terms.

Neither of these two categories of needs - the purely physical prerequisites of survival or needs further up on the scale of wants - is necessarily reflected by the economic measures at our disposal. These measures - like the majority of forecasts - are based on market transactions between sellers and buyers. One condition that has to be met in order for a transaction to materialize is that the buyer must have purchasing power. Poor people have by definition not enough money to make the purchases which they desire to make, and consequently their actual needs will not be reflected by the market or by statistics. A certain growth or level of GNP, therefore, tells us nothing concerning, say, the standard of nutrition in a country or the strength of pent-up desires for an improvement in living standards.

The GNP concept has lately been subjected to a critical scrutiny which, however, we have no cause to consider in detail for present purposes. An American economist, for example, has argued that an increasing proportion of GNP - a portion over and above basic needs - is mainly concerned with goods or services whose true value consists in their very scarcity, i.e. in only a few people having access to them. The greater the number of people to whom these goods (e.g cars) become available, the less pleasure the owners will derive from them (due, for example, to traffic jams).(1) In any case, we may observe that the economic measures which are used to reflect demand in markets do not tell us whether the needs provided for are of greater or lesser importance.

Then again, the concept of needs changes with the passing of time. It is not all certain - indeed, it is quite unlikely - that people in twenty-five years' time will view their needs in the same light as we view ours today. An incipient or imminent shortage of raw materials may accelerate the advent of a society more economical in its use of resources. But even without such a prospect, patterns of consumption, even patterns of living, are likely to change. The best way of appreciating this is perhaps to look back on developments in e.g. Sweden from the early 1950s onwards and to make comparisons with the present day situation as regards the number of cars, cinema visits, working hours and holidays, the number of day nurseries, the things bought in food shops, the size of the public sector and so on. The pattern of developments in Sweden is more or less the same as has occurred in other affluent countries. In the developing countries too, the post-war era has witnessed a change in views as to which commodities are necessary.

So far, very little research has been done concerning the amount of raw materials that would be needed in order to assure everybody of a certain living standard. In fact a great deal of research remains to be done before we can even begin to answer such questions.

It is therefore impossible today to give a clear-cut answer concerning the amount of raw materials that will be 'needed'. Instead the point of departure for the following argument is the <u>demand</u> which may appear in the market, and the majority of forecasts actually refer to demand and not to the satisfaction of needs.

Forecasting demand for raw materials

The economic expansion of the past hundred years has brought a rapid increase in the consumption of raw materials. Figures 22-25 show that annual output of raw materials - which in a long-term perspective is the same thing as consumption - in many cases multiplied a hundred times between 1870 and 1960. Developments during the 1960s - which are not included in the figures - brought even swifter growth in many cases. But there are many other instances where production has grown at

(1) Fred Hirsch: 'Social Limits to Growth', Routledge and Kegan Paul Ltd., London 1977.

a fairly steady annual rate, and the commonest method of forecasting developments concerning individual raw materials is that of trend projection. For example, if consumption of wheat and steel respectively has risen by 2.5 per cent and over 4 per cent per annum over the past twenty or forty years, one assumes that it will continue to do so in future.

Closer inspection of consumption figures reveals that raw material consumption most often varies parallel to total production, as measured by GNP. The link between GNP and consumption of various raw materials often applies over long periods of time, but it also holds fairly true during cyclical fluctuations, for the simple reason that when industrial production slows down or speeds up, industrial consumption of raw materials does likewise. But a closer analysis of the short-term often reveals lags in time, because raw material inventories also change in the course of the business cycle. This simple connection between GNP and raw material consumption is therefore employed in a large number of forecasts. The technique is to estimate the amount by which raw material consumption rises for every one per cent of GNP growth and then to make an assumption concerning the future growth of GNP.

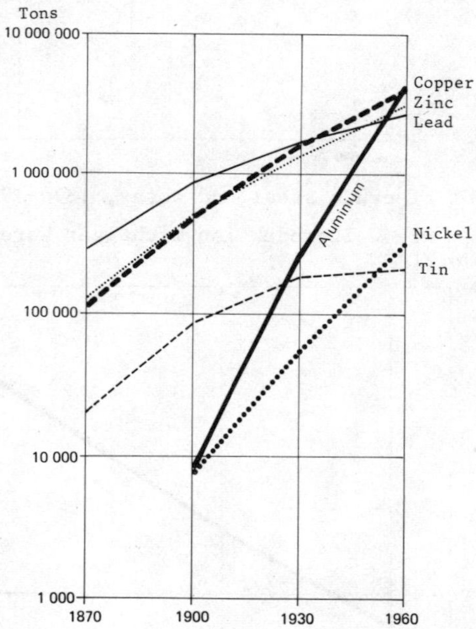

Figure 22. World output of the six most important non-ferrous metals, 1870-1960.

Note: The 'knees' in the graph lines are merely due to the drawing technique, which involves drawing straight lines between the few points which are given.

Source: W. Vigener, "Die Weltproduktion wichtiger Waren", Berlin 1970.

One criticism levelled against this method is that it fails to take into account differences which may be due to differences in the development levels of different countries and changes occurring in this respect with the passing of time. Nor does it make any allowance for the changing composition of output and consumption. The fact that the more affluent countries use an increasing proportion of their income on services ought reasonably to entail a gradual decline in the consumption of materials associated with a given GNP increase. Figure 26 illustrates this with reference to steel. More recent studies, therefore have begun to use the concept of material intensity, which refers to the quantity of raw materials

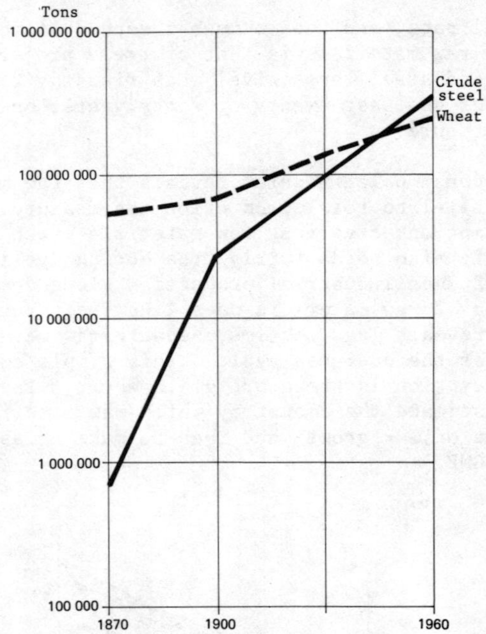

Figure 23. World output of crude steel and wheat, 1870-1960

Source: W. Vigener, "Die Weltproduktion wichtiger Waren", Berlin 1970.

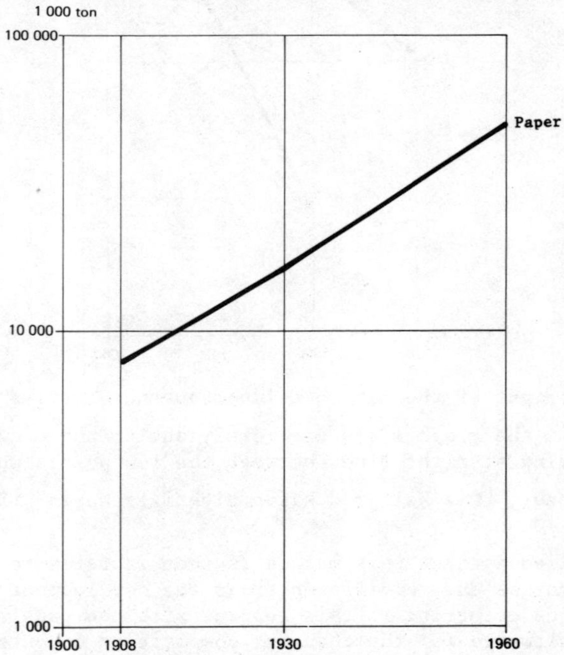

Figure 24. World output of paper, 1908-1960

Source: W. Vigener, "Die Weltproduktion wichtiger Waren", Berlin 1970

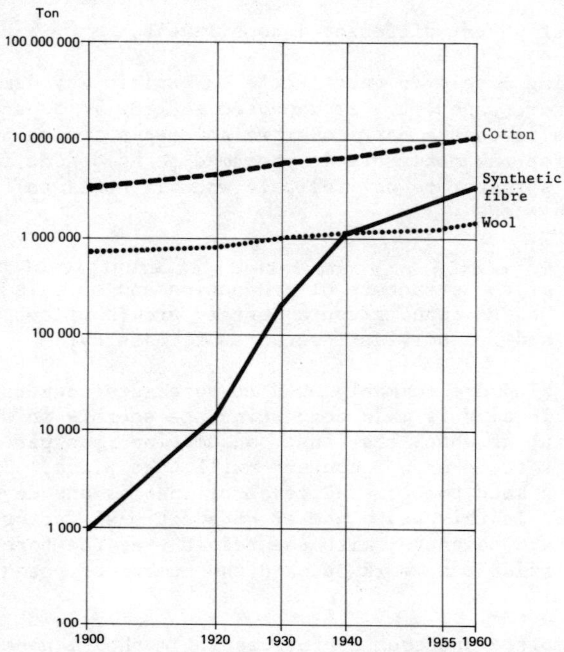

Figure 25. World output of the principle textile fibre materials, 1900-1960
Source: W. Vigener, "Die Weltproduktion wichtiger Waren", Berlin 1970.

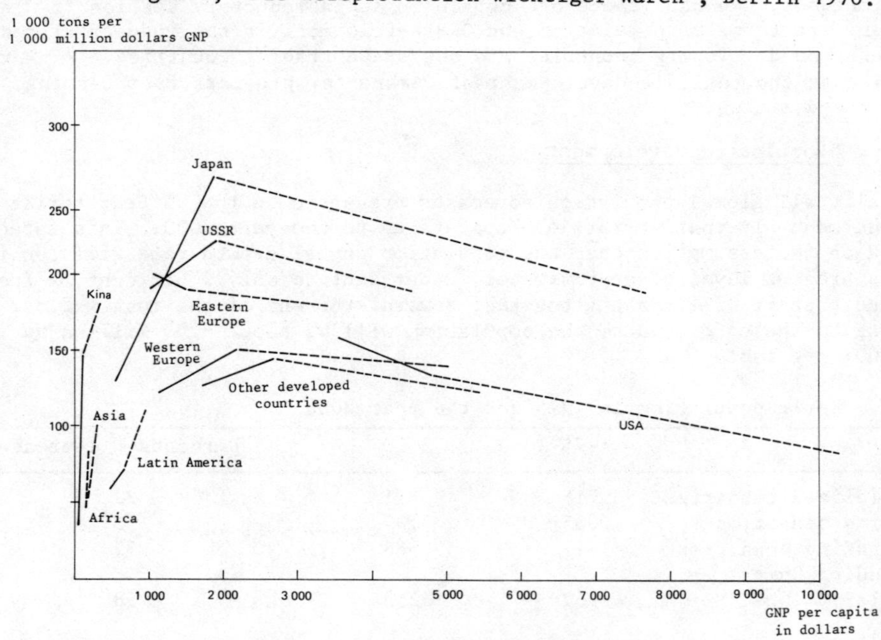

Figure 26. Material intensity at different levels of GNP. Steel
Source: Wharton School of Finance (project leader W. Malenbaum): "Material Requirements in the US and Abroad in the Year 2000", Pennsylvania USA, March 1973.

consumed per unit of GNP at different income levels.

But these forecasting models do very little to explain why demand for steel, artificial fertilizer, paper etc. is expected to grow in a particular way. All the method tells us is that a particular relationship in the past between total output and demand for raw materials is presumed to hold good for the future as well. The method is therefore not reliable when it comes to finding out what is actually going to happen.

There are a number of models on a completely different level of reliability which describe the whole of the structure of production and on this basis indicate which raw materials will be in demand given a certain growth of output in both the economy as a whole and in individual sectors of industry.

Models of the same kind are commonly used in forecasts concerning individual raw materials. An inquiry is made concerning the sectors in which the raw material is used and the point at which its final consumption - as part of a motor vehicle, a washing machine, a train or a computer - will take place. Exact assumptions are then made concerning both the general level of incomes and developments in individual sectors. In this way a number of additional factors are included which are intrinsically hard to gauge, with the result that the forecast becomes technically more difficult to work out and the number of potential sources of error is increased.

Thus there is a complete spectrum of forecasting methods, none of which is dead certain, least of all when looking as much as twenty-five years ahead. One thing is clear, however; general economic developments play an important part in all the various methods. We have therefore chosen in an introductory section to discuss conceivable patterns of population and GNP development in the industrialized countries, the developing countries and the state trading countries.[(1)] Next we shall discuss the possible development of demand for products from farming, forestry fisheries and mining.

Population development

Practically all global population forecasts are based on the UN Secretariat's so-called mean alternative for developments up to the year 2000. This forecast is based on the assumption that the population annual growth rate will decline from its present level of approximately 2 per cent to only 1 per cent as from the 2010, and that it will cease altogether towards the end of the twenty-first century. According to these forecasts, the population will be about 6200 million by the year 2000; see table 3.

Table 3. World population in 1975 and the year 2000

	1975	2000	Percentage increase
Industrialized countries	715	871	22
Developing countries	2055	3797	83
State trading countries (including China)	1197	1585	32
The world	3967	6253	58

Source: UN, cit. Population Reference Bureau Inc.

(1) The three groups of countries correspond to those commonly employed in United Nation's analyses. The term "state-trading country" is used synonymously with "centrally planned" according to UN vocabulary.

The world's population is thus expected to rise by almost 60 per cent during the next twenty-five years. Almost 80 per cent of this increase - about 1700 million people - is expected to occur in the developing countries.

The forecast is interesting from at least three points of view. Firstly, it is lower than previous forecasts. In fact this already applies to figures for 1975, which fall short of previous assessments by 55 million. The forecast for the year 2000 is 240 million lower than the previous forecast. There are some indications that fertility has declined faster than had been expected.

Secondly, the forecast for the year 2000 is relatively insensitive to different assumptions concerning birth rates. The population structure of the world as a whole, but particularly that of the developing countries, is such that very large cohorts will reach child bearing age during the next few decades.

However different assumptions on the pace of change in birth rates figures have long-term implications for the stabilized level of size of population in the distant future. The future population of Mexico can be taken as an example of this phenomenon. The population of that country is now about 60 millions, and it is growing at a rate of some 3.5 per cent annually. If the birth rate were to decline so drastically that as early as 1980 or 1985 every woman had only one daughter who survived to marriagable age, the growth of population would still not cease until the year 2050, by which time the country would have 110 million inhabitants. If instead we assume that the birth rate reaches this level in the year 2000, the population will be stabilized at 175 million.

Thirdly, the table shows that even the slow population growth in the affluent countries (barely one per cent per annum) will in the course of twenty-five years produce a substantial increase in absolute figures (over 150 million people, which is more than half the present day population of the EEC countries).

For all their brevity, the above remarks on population growth require some elucidation. Forecasts, even when they refer to developments over such long periods as 25 or 50 years, are often stated to the nearest million (in Sweden to the nearest thousand). This mirrors in no way the still great uncertainty particularly regarding the statistical base. The fact is that we do not know even to the nearest hundred million - perhaps to the nearest 400 million - how many people are living today in this world. It has often happened that a new census in one of the developing countries has made the headlines because its results have widely deviated from previous suppositions. Not even estimates of the populations of highly industrialized countries are altogether free from elements of uncertainty.

GNP growth

It is not easy to forecast the growth of total production, even if we confine ourselves to the immediate future.

The 1970s have shown that the most sophisticated methods of economics and statistics have not foreseen the very rapid boom in 1973, the duration and depth of the deepest recession so far during the post-war era, or the timing of the 'real' recovery. One can of course plead that developments have been influenced by factors which no forecast could take into account, particularly the economic policy actually pursued by the affluent countries - which is not necessarily the same thing as the policy officially stated by their governments.

The difficulties involved in sizing up developments within a short period of time become correspondingly greater if the perspective is extended to 25 or even to as little as 10 years ahead. There are theories of long-term economic growth which are also used in estimates of various kinds concerning possible development, but these serve mainly as a foundation for discussions about the future and are less frequently used as a basis for forecasting.

Discussions of long-term economic development are therefore mainly confined to the projection of trends, sometimes coupled with arguments as to why developments may deviate somewhat from the trend. Occasionally these arguments are based on theorectical grounds, but in all cases the underlying assumptions - i.e. those determining the magnitude of GNP growth - are fairly rudimentary. The following extract from a World Bank forecast is typical: "for the period 1981-85, economic growth in the developed countries has been projected at 4.9 per cent per annum, in line with developments between 1961 and 1973". The FAO forecast prior to the World Food Conference in 1974 presupposed that the economic growth of the developed countries would be inhibited by the energy crisis, but that this would still not prevent growth from amounting to 4.1 per cent annually between 1970 and 1975 and then rising to 4.6 per cent during the second half of the 1970s and 4.7 per cent during the 1980s,"compared with an average of 5.0 per cent during the sixties". Of course, this adjustment downwards in relation to the trend during the 1960s represents a purely discretionary assumption. The simple fact is that there do not exist any reliable methods of predicting long-term economic development. What is more, the subjective uncertainty of such predictions is now greater than ever because the prolonged recession has broken the trend.

However, it is probably possible to say something concerning the limits to possible development. The lowest alternative forecast which we have found[1] indicates a per capita GNP growth of 0.4 per cent per annum up to the year 2000. Given an annual population rise of barely 2 per cent, this corresponds to a world GNP increase of between 2 and 2.5 per cent. There is reason to query the compatibility of such a low growth rate with a world economy functioning on the present lines, and a forecast of this kind should therefore also include drastic structural changes to the world economy. One of the first objections to be made is that such a long, continuous period of very small growth has not occurred in the industrialized countries since 1870. The relative social stability of the present day industrialized market economies is considered by many people to be in important respects a result of rising GNP. In addition, there have been considerable scope for technical innovation.

An annual GNP growth rate of about 3 per cent, which implies a doubling of world GNP in twenty-five years, is therefore probably to be regarded as an absolute minimum alternative when discussing a world similar to that of today. On the other hand, a 6 per cent annual growth rate, i.e. a quadrupling of world GNP in twenty-five years, would exceed that of the early post-war period, which is the highest in modern times.

Most forecasts come within these limits, and practically all of them presuppose that growth in the developing countries will exceed that in the industrialized countries. This is not unnatural, partly because the developing countries should in principle have far greater opportunities of productively occupying their unemployed or under-employed populations. Nor does a lower growth rate for the industrialized countries seem unlikely, for example, in view of the tendency for more and more people to be employed in the service sector, where productivity gains, as they are measured today, are lower than those of industry, for instance.

We can now illustrate the conceivable course of developments by means of two 'forecasts'. One of them is more of a trend projection and comes from the forecast

(1) G J S Govett and M H Govett, World Mineral Supplies - Assessment and Perspective, New York 1976.

drawn up for the World Food Conference in 1974. This forecast really only extends as far as 1990, but it may be presumed to reflect a general assessment of conceivable long-term growth. 1990 is only considered as one of the points on a national growth graph. We have used the forecast as a basis on which to calculate developments for the remainder of the present century, on the assumption that the trends from the 1960s continue to hold.

The other forecast comes from a study compiled for the United Nations by the Russian-American Nobel laureate in economics, Wasilly Leontief.(1) This forecast presents a number of alternative calculations, and we have chosen the alternative where the objective is a reduction of the per capita income gap between industrialized and developing countries from 1:12 (its present size) to 1:8 by the year 2000. This forecast, based on a clearly specified objective (and thus employing what is termed a normative approach)(2), can be said to constitute an extreme case among the forecasts we have studied, both as regards its target for growth in the developing countries and as regards the difference between the growth rates of industrialized and developing countries.

Leontief assumes that world GNP growth will be slightly lower than the trend (see table 4), since he has deliberately reduced the growth of the affluent countries.

Table 4. Annual growth of GNP according to the 'trend' and according to the Leontief study

	Percentage annual GNP growth	
	Trend	Leontief
	1970-2000	1970-2000
Industrialized countries	4,5	3,8
Developing countries	6,4	7,4
State trading countries	5,9	5,4
The World	5,1	4,8

If, using the above mentioned UN forecast for the year 2000, we convert these figures into GNP per capita, the normative element in Leontief's forecast becomes even more apparent; see table 5.

Table 5. GNP per capita

	GNP per capita, dollars in fixed prices		
	1970	2000	
		Trend*	Leontief*
Industrialized countries	2850	8300	6825
Developing countries	220	650	870
State trading countries	635	2400	2100
The World	890	2250	2070

* GNP in the year 2000, given the growth rates assumed in table 4.

(1) W Leontief et al., The Future of World Economy, UN New York 1977. (The reference here is to scenario ten).

(2)'Normative' implies assumptions according to what is considered desirable on the basis of stated values, rather than assumptions based on the likeliest course of development.

The difference between the two forecasts corresponds to the whole of the per capita output of the developing countries in 1970, or one-third of their output in the year 2000. What is even more striking is that the difference between industrialized and developing countries will continue to grow rapidly in absolute figures. In 1970 that difference was about 2600 dollars per capita and, according to Leontief, it will be almost 6000 dollars by the year 2000. The trend forecast points to a difference of some 7600 dollars.

Thus not even Leontief's highly optimistic assumptions concerning economic growth of the developing countries point to the achievement, in absolute terms, of a more equitable world during this century. Nor can the developing countries be expected to account for a share of world production which bears any reasonable resemblance to their share of world population. Although their economies are expected to grow faster thus accounting for a growing proportion of world output, the industrialized countries will continue to dominate production and raw material consumption with it; see table 6.

Table 6. World GNP in 1970 and in the year 2000

	1970	2000	
		Trend	Leontief
Industrialized countries	2100	7900	6460
Developing countries	380	2450	3290
State trading countries	734	4100	3510
The World	3222	14450	13260

At the same time it is a sobering thought that current estimates covering a twenty-five-year period actually point to an output four times greater than today's, even in real terms, i.e. without figures being inflated by rising prices.

There is some justification for arguing that the forecasts presented here mark the upper limits of probability. More recently there has been a tendency to tone down assumptions regarding long-term economic growth, as witness the forecast employed by the US National Commission on Materials Policy[(1)], whose assumptions concerning per capita GNP growth are reproduced in table 7.

Table 7. Assumptions made by the US National Commission on Materials Policy concerning growth

	Percentage annual increases 1971-2000		
	GNP per capita	Population	GNP
OECD countries	2,75	1,1	3,8
USA	2,5	1,2	3,7
Developing countries	1,5	2,4	3,9
Eastern Europe	2,5	1,3	3,8
China	2,9	1,4	4,3
World	2,0	1,8	3,8

(1) National Commission on Materials Policy, Washington 1974.

Whereas previous forecasts arrived at a world GNP growth of about 5 per cent, this forecast is more than one percentage unit lower. The growth rates for all groups of countries are lower. The growth rates for all groups of countries are lower than before, but the most striking difference concerns developing and state trading countries; their figures deviate from the trend forecast by more than 2 per cent.

It is significant that the main focus of uncertainty is on such a group of countries for which even extreme opinions can still be respected. The figures for the OECD area, which is under close economic surveillance and where many institutions have clear-cut views, do not deviate from 'the trend' by more than 0.7 per cent per annum. Of course, not even this difference is negligible, for in the course of 10-25 years it will produce a GNP of quite a different magnitude. Clearly, however, the uncertainty surrounding economic developments is substantially greater than the differences between forecasts would suggest.

Even assumptions of a more extreme nature cannot be dismissed as completely impossible. We may not be able to maintain anything like the same growth level as previously. At the end of the 1960s, who could ever have conceived of the 1970s in terms of stagnant industrial output, 10 per cent inflation, heavy unemployment and, for a few years at least, falling real wages as a problem affecting more and more countries? It is not easy to condense into a short summary all the various explanations that have been proffered as to why this actually happened. One short-term, partial explanation is that normal cyclic fluctuations have been aggravated by exogenous circumstances i.e. factors of an outward kind like the oil crisis or rising international uncertainties. Attempts at more long-term explanations range from the growth of the public sector and high taxation on the one hand to diminishing profit ratios and the crisis of capitalism on the other.

One thing is clear. The present recession is the longest we have had since the 'great' crisis of the 1930s, and each new forecast makes it look like growing longer and longer.

We do not believe that a great economic depression or crisis is the most probable course of events, but the possibility cannot be discounted, and if it were to materialize, social stability and world trade would come under a severe strain.

But what will happen is not predetermined. When, how and if the present crisis will be resolved is a question for the economic policy of governments. In a more long-term perspective, the method by which these problems are tackled will have crucial bearing on both the level and the geographical distribution of growth.

Demand for raw materials

Agriculture

How would demand for foodstuffs be affected by the developments outlined above?

We may presume that demand for foodstuffs in the affluent countries is not going to rise very drastically. Although the situation varies from one country or income group to another, the industrialized countries as a whole are approaching saturation level, at least where calory intake is concerned. Rising income cannot be expected to have more than a marginal effect on consumption. It has been estimated that a one per cent rise in income in these countries will produce an increase of barely 0.1 per cent in demand for calories for direct human consumption. Instead the rise in income is used for the purchase of more meat. This in turn boosts demand for fodder grain among other things, and total demand for grain can therefore be expected to rise somewhat more in relation to income.

In poor countries, however, it is to be expected that a greater proportion of increases in income will be used for purchases of food, expressed both in calories and in protein content.

Given the current distribution of income in the developing countries, it has been estimated that a rise in income will lead to a growth in food demand which is more than twice that occurring in the industrialized countries. The difference would be even greater if it were not for the existing distribution of incomes.

We have already discussed the difference between demand and needs, and there is no more glaring example of this difference than the food situation of the developing countries. The forecasts compiled prior to the World Food Conference envisaged a 3.5 per cent annual increase in food demand in the developing countries for the period ending in 1985. But at the same time the number of severely undernourished people in the world was expected to rise from about 500 million to over 800 million. In other words, the difference between purchasing power and actual food requirements in calories was expected to increase.

The uneven distribution of incomes in the developing countries is one of the fundamental reasons why needs are not manifested in the market. Therefore a general rise in incomes is of minor importance as an explanation for food demand. An OECD study[1] has shown that calory demand can rise by 0.3 per cent annually given a total annual income rise of 4 per cent, if the whole of that increase goes to the richest one-fifth of the population of a developing country. But the rise in demand will be no less than 3.2 per cent if the same addition is made to the incomes of the poorest two-thirds of the population. A redistribution of income in favour of the poorest groups also causes food demand to rise, even if total incomes remain unchanged. A serious forecast of the development of demand therefore requires a presentation of assumptions concerning the internal distribution policies of the developing countries.

Similarly, forecasts of demand in the affluent countries should take into account which products will be subsidized, because the agricultural policy pursued hitherto has directly and palpably influenced the pattern of consumption.

Distribution policy, however, has so far not had any direct bearing on global forecasts. Reference is often made to the desirability of internal reforms to establish equity, but forecasts are based on the present-day distribution of incomes.

What do these forecasts tell us? In 1974 the FAO predicted that demand in the industrialized countries would rise by 1.5 per cent annually as against 3.6 per cent in the developing countries. Demand in the world as a whole was expected to rise by 2.4 per cent, 2 per cent resulting from population growth and the remaining 0.4 per cent from increased purchasing power.

Demand for meat was expected to rise by 3.1 per cent per annum and demand for cheese by 2.8 per cent. Grain and root crops were expected to rise by 2.4 and 1.4 per cent per annum respectively.

Leontief's normative forecasts are consistently higher. He expects demand for grain to rise by nearly 4 per cent in the world as a whole, and by 5.1 per cent in the developing countries. Demand for animal products is expected to rise by 3.4

(1) OECD, "Study of trends in world supply and demand of major agricultural commodities", p. 43.

per cent, the developing countries accounting for more than half of this increase.

The difference between a trend forecast and Leontief's more normative approach will be clearly apparent from the following comparison of demand for grain in the world as a whole and in the developing countries in the year 2000 as compared to the situation in 1970.

Table 8. Demand for grain in 1970 and in the year 2000, according to 'the trend' and according to the Leontief study

	Grain demand, millions of tons		
	1970	2000	
		Trend*	Leontief
Developing countries	386	1020	1625
Other countries	831	1380	1846
The World	1217	2400	3471

* The FAO forecast has been extrapolated to the year 2000 for the sake of comparison.

The normative aspect of the forecast is most apparent from the supposition that the developing countries will be able to more than quadruple their grain demand. It should be observed,however, that even at this level the developing countries will still fall short of the volume of demand in the affluent western countries in 1970. (The "Other countries" group in table 8 also includes such countries as China).

At the same time Leontief's forecast seems dubiously high regarding grain demand in the other countries, above all because he expects their incomes to be lower than those envisaged in the trend forecast.

Fisheries

Our two forecasts view fish consumption from completely different perspectives. Prior to the World Food Conference of 1974, the FAO estimated that demand for fish for human consumption would rise from about 40 million tons in 1970 to about 80 million tons in 1990, i.e. by 3.4 per cent annually. Leontief, on the other hand, has assumed that catches will remain at the 1970 level and that the proportion of industrial fish - i.e. the proportion used mainly for livestock feed and not for human consumption - will remain at 25 per cent.

Both these statements should be taken with a pinch of salt. In the first place, they are both influenced by the fact that stocks of traditional species are already being overfished in many regions. Thus allowance has already been made for limitations of supply. But there are quite considerable possibilities of increasing catches of non-traditional species such as cuttle-fish and krill. The forecasts make no allowance for these possibilities. But demand for fish is not certain to rise with income. It may very well do so where certain species are concerned; people in high income classes eat larger quantities of such 'exclusive' fish as salmon and sole. The time may soon come when herring is also counted as 'high class' fish, but most species count as everyday food. In cases of the latter kind, rising income causes consumers to go in, say, for sausages instead. Both within and between different countries, patterns of reaction to rising income can be expected to vary. In our sub-study entitled 'The Sea and Marine Resources', Hans Ackefors had the following to say on this subject.

"People in Japan eat cuttle-fish, fish sausage and marine algae. People in Sweden do not. Our social and cultural heritage governs our eating habits and the values we put on different fish species as food. People in Norway prefer saithe to cod, and in Sweden the opposite applies. The Danes relish dried salt cod, but the Swedes are more partial to dried ling or cod. Great amounts of valuable fish are rejected in different parts of the world because they do not fit into traditional eating habits. The superior or inferior nutritional value of the fish itself does not made any difference either way.

Cuttle-fish is greatly in demand in certain parts of the world such as Japan and the Mediterranean countries. But there are other large parts of the world where it is unsaleable. Sweden's landings of cuttle-fish from the Skagerrak during the 1940s and 1950s could only be sold in Southern Europe, and nowadays Swedish fishermen no longer fish cuttle-fish. It is possible, however, that people in the affluent countries will have to revise their attitudes at a future date".

If people in the affluent countries do revise their attitudes, perhaps demand will be substantially greater than the forecasts suggest, but this additional demand will mainly centre on fish novel to human consumption.

Forestry

Table 9 shows the pattern of consumption for three groups of countries in 1973. A discussion of the future development of demand is bound to take as its starting point the great differences between industrialized and developing countries as regards the use made of wood.

Table 9. Consumption of forest products

	Total		Per capita			Population
	Sawn timber million m^3 (s)	Paper products m. tons	Firewood m^3	Sawn timber $m^{(3)}$ (s)	Paper products kg	millions
Industrialized countries	253	119	86	340	160	745
Developing countries	37	11	444	20	6	1879
State trading countries	154	17	207	(127)	(14)	1211
Total	443	148	299	116	39	3824

Source: "Renewable Raw Materials", the RoR study, report 304, 1976, p.65. (In Swedish). The figures are based on FAO data from 1975. (s) means 'solid'.

Forest industry products are dominantly consumed by the industrialized countries. The western industrialized countries account for 80 per cent of world consumption of paper products. The developing countries account for 12 per cent of sawn timber consumption, 6 per cent of board product consumption and 11 per cent of the consumption of paper products.

The industrialized countries consume 17 times as much sawn timber per capita as do the developing countries, 52 times as much board and 27 times the amount of paper products. The average Swede consumes about 25 times more industrial timber than the average inhabitant of a developing country.

Total consumption of wood per inhabitant, however, is only about three times greater in the industrialized countries than in the developing countries. The reason why this difference is not greater is that the developing countries use such a large proportion of their timber as fuel.

This latter observation has an important bearing on future proposects; 45 per cent of the raw timber is used today as firewood, mostly in the developing countries. This demand is only manifest to a limited extent in the market. At the same time it is a fairly safe bet that demand for firewood will diminish with rising income. Very simply, people switch to oil derivates such as kerosene in the first instance and then to electricity. The reverse applies to forest products - sawn timber, board products (e.g. veneer) and paper - because demand for them will expand with rising incomes.

The FAO forecasts for the 1980s envisage a 2.9 per cent rise in world demand for raw timber. Consumption of firewood is expected to rise by about 2.5 per cent, which more or less corresponds to population growth in the developing countries. If these growth rates are applied to the period ending in the year 2000, demand for raw timber will then amount to 5400 million m^3 as against some 2500 today; in other words, it will more than double in the space of a quarter of a century. Leontief gives much higher growth rates for timber goods and paper products. In round figures he discusses an increase totalling between three and four times the present volume of consumption, with annual growth rates of between 4.5 and 5 per cent. He expects increases in the developing countries to exceed 7 per cent. Although demand in the affluent countries is expected to rise at about half this rate, the increase for these countries will account for between two-thirds and three-quarters of the global increase, because it starts from a much higher level.

In the RoR study "Renewable Raw Materials", it has been assumed that consumption of industrial raw timber will rise to about 3000 million m^3 by the year 2000, compared with 1400 million m^3 in 1973; and consumption of firewood, to 1700 or 2000 million m^3 (as against 1000 million m^3 in 1973).

Minerals

The RoR study "Non-Renewable Raw Materials" reviews a number of forecasts for the mineral sector, but we shall confine ourselves to the latest forecasts of the US Bureau of Mines and to those made by Leontief: see table 10.

Table 10. Two forecasts of metal consumption in the year 2000

	Consumption in 1974,1000 tons	Annual percentage increase 1930-74	Annual consumption growth rates, 1970-2000		Consumption in the year 2000 as forecast by Leontief (1000 tons)
			US Bureau of Mines	Leontief	
Steel	513500	4,3	2,8	5,2	1920000
Nickel	704	6,1	2,6	4,8	2382
Lead	3701	2,1	2,3	5,5	14890
Copper	7257	3,7	3,9	5,1	26450
Zinc	5806	3,4	2,3	4,7	19164
Aluminium	15420	9,4	5,2	4,8	52177
GNP increase			3,8	4,8	

As can be seen, the differences are very great. Probably this can be partly attributed to the very high economic growth rates for the developing countries assumed by Leontief and what is perhaps more important still, to his assumption that economic growth will be so much greater in the developing countries compared to the industrialized countries. This may serve to explain part of the higher growth for steel, for example. One general point of criticism concerning Leontief's forecasts for individual raw materials is that insufficient regard is paid to shifts between raw materials.

The growth rates indicated by the American forecast are very low, substantially lower than the trends. This is at least partially explained by the assumption of a relatively lower GNP growth. The fact that material intensity is greatest at a per capita GNP of 2000 dollars is probably still more important. Because of their slow growth, the developing countries as a group will approach this level very slowly, which means that demand for raw materials will also grow slowly. Mineral intensity in the affluent countries, on the other hand, is presumed to decline when GNP rises, which means that the growth of raw material consumption in these countries will fall short of their assumed GNP growth.

Summary

Material intensity changes with rising incomes and with technological progress. Income distribution within countries will do a great deal to decide which raw materials are going to be in demand. This is most palpably true of the food sector. A perpetuation of the present uneven distribution of incomes, especially in the developing countries, will here result in a rising demand for meat. A greater equalization of incomes would instead bring about a steeper rise in demand for staple food products (grain, root vegetables and milk).

Of course, there are other factors involved as well. Increased recycling may reduce demand for virgin raw materials. Technological progress may reduce the need for raw materials to perform a certain function or to produce a commodity. European cars contain practically the same raw materials as American cars, but on average they weigh just over half as much. A modern electric motor with an output of 1kW weighs roughly half as much as its precursor from the 1930s. An American car battery wieghed more less the same in 1960 as it had done in 1940, but its service life was 50 per cent longer.

Finally one cannot exclude the possibility of demands for a better occupational environment and protection of the external environment imposing restrictions on demand for individual raw materials. Examples of the latter kind are provided by the ban on DDT in several countries, the decline in the use of mercury and demands for a reduction of the lead content of petrol.

It is very hard to size up the potential effects of these factors on material intensity, but since they are changing so slowly, economic development remains the central consideration regarding the twenty or twenty-five-year perspective with which we are concerned. We may therefore conclude with the following observations.

Overall demand for raw materials will hinge on future economic growth. It is true that there are tendencies in the affluent countries for heavy and/or expensive raw materials to be replaced by others, for the material content of products to be reduced and - what is still more important - for services to be more extensively demanded instead of goods. But these factors carry so little weight in the context of twenty-five years that demand for raw materials will rise if the world's GNP continues to increase at predictable growth rates, i.e. by between 3 and 6 per cent annually.

The economic growth of the developing countries has an important bearing on demand for raw materials. If one assumes (and/or hopes) that they will come to account for a growing proportion of the world's GNP, demand for steel, paper, tungsten and rubber, for example, will grow more rapidly than if their share of GNP were to remain stationary.

Even if we assume that growth in the developing countries will be substantially faster than in the industrialized countries, the latter will dominate world production and raw material consumption for the coming twenty-five years, provided that the global economic system then bears some resemblance to that which we have today.

4 Global Supply

In this chapter we shall discuss the way in which the supply of raw materials may possibly develop. Gradually and in the long term, supply must of course develop in parallel to realized demand. It is inconceivable, for example, that output should rise by 3 per cent and consumption by 2 per cent in the course of ten or fifteen years, and indeed hardly for a period of a few years either. Otherwise vast stocks would accumulate. Nor can production, expressed in tons, be expected to rise more slowly than consumption. It may do so for a few years, while surplus stocks are worked off, but not for a decade. If, therefore, potential demand exceeds potential supply, prices will rise until realized demand, i.e. the amount actually bought, equals supply. A great deal of economic debate - and forecasting - in the raw materials sector is in fact concerned with the future development of prices.

We shall not be venturing into this difficult territory. Instead our purpose in this chapter will be to discuss how the supply of raw materials can be expected to develop in the light of the results presented in some of our interim reports on the limits to raw material production. The limits considered are physical and social. After this summary, which looks at individual sectors, consideration will be given to a number of factors common to all raw material sectors. The conclusions we present, in the form of figures, serve two purposes. First they are intended to convey an idea of the magnitudes of current assessments in international debate, and secondly they are intended to point to the wide variety of those assessments. It should be emphasized that we are above all referring here to the next twenty-five years; a more long-term perspective is considered in Chapter 5.

Agriculture

It is very hard to make a collective assessment of the physical and technical possibilities of raising world output of food, especially where grain is concerned. And it is even more difficult to predict what is actually going to happen, i.e. to take account of the social factors influencing production as well as the purely technological conditions.

Most of the world's experts agree, however, that food production can be multiplied several times over. The most optimistic of them believe that the long-term upper limit is that which would be achieved if the present standard of the industrialize

Irrigation is one of the prerequisites of a growth of food supply. This picture shows a farmer in Pakistan cleaning a channel in the intricate irrigation network so that the water will be spread more efficiently over the cultivated land. (Picture: World Bank and IDA)

countries were applied to feeding between 30000 and 60000 million people.[1] Although the level attainable in practice is probably far below this, few scientists would query the technical possibilities of substantially improving output by measures in the following sectors.

1. Additional acreage. Only 10 per cent of the earth's land area is under the plough. The cultivated acreage can be increased somewhat, but the scope for such an increase is unevenly distributed. This is especially true in the developing countries, where the need is greatest, but the FAO has set a general target for the developing countries, providing for an increase of some 20 per cent in their cultivated acreage between 1970 and 1985.

2. Improved yields. Crop yields per unit of area are also very unevenly distributed among the countries of the world. One hectare of cultivated land in Sweden yields more than three times as much grain as an average hectare in the developing countries. If all countries were to attain yields corresponding to Sweden's, the world's output of grain would rise by 150 per cent. Clearly there is much more to be gained from improving the yields of existing acreage than from expanding the acreage itself.

What is behind the difference in yields? The most general answer to this question is: the extent to which agriculture employs productive resources from other sectors of the economy. High yield farming is a more or less industrial concern involving massive inputs of artifical fertilizer, machinery, energy, pesticides, etc. But there is no simple formula for increasing the supply of these resources to low yield farming, because it is social, economic and political factors that decide whether they will be used and, if so, how they will be used and to what purpose. If these conditions made it possible for such inputs to be deployed efficiently, and if we assume that yield per hectare is roughly proportional to fertilizer dosage, the main requirement in order to double the grain output of the developing countries would be to raise average fertilizer dosage by more than 350 per cent. Needless to say, this would entail very great expense - something in the region of 10 billion US $ per annum.

Because artificial fertilizer is an expensive input, attention has been drawn to the inefficient use that is made of it at the global level. Diminishing marginal yields at higher levels imply that the last few kilos of artifical fertilizer per hectare in the industrialized countries give a smaller return than the first few kilos of artificial fertilizer could produce in the developing countries.

Figures which have emerged from Swedish experiments can be used in the following hypothetical example. If Western Europe were to reduce its consumption of artificial fertilizer from 200 kg per hectare to 175 kg per hectare, this would reduce grain output by about 10 million tons. If the artificial fertilizer thus saved were to be used in Africa instead (at a rate of between 5 and 18 kg per hectare), grain production there could be raised by 17 million tons. This would give a total gain of 7 million tons, which is more than the whole of Sweden's output. This, however, is a purely hypothetical example.

Increased irrigation is one means of raising yields and also, in practice, of expanding the cultivated acreage, because it sometimes means that up to three crops can be harvested in a year. The FAO believes that the irrigated acreage of the developing countries could be raised from 13 to 18 per cent of the total area under cultivation. But an expansion of this kind requires a great deal of capital and has to be balanced against other possible improvements.

(1) "Renewable Raw Materials", the RoR study, report 304, 1976, p.35 (In Swedish)

The Green Revolution is a collective term which has been applied to measures in connection with the introduction of new high yield varieties of wheat, rice and maize, measures which have also demanded the stepping up of other inputs such as artificial fertilizer, irrigation and pesticides. Few programmes for improving world food production have so distinctly underlined the differences between what is technically feasible and what is socially possible in the agricultural context.

A review of the various technical preconditions makes it perfectly clear that the potential scope for increasing agricultural output is so great that it cannot possibly be fully utilized within the next twenty-five years. This is partly due to economic reasons. The FAO has estimated that a programme aimed at expanding agricultural acreage, improving existing irrigation facilities and enlarging the irrigated acreage in the developing countries would cost about 80-100 billion US dollars spread out over a ten-year period. In Southern Asia, also according to the FAO, current expenditure on improved seed, commercial fertilizer and pesticides would have to be raised from US dollars 5 to 15-60 per hectare. Simplifying matters, if we assume that average expenditure in the developing countries had to be raised by 10 US dollars per hectare this would mean a total annual increase of something like 13 billion dollars. Altogether, therefore, agricultural development would require an increase in the available resources by something like 20 billion US dollars per annum, which is about 3 per cent of the GNP of the developing countries and about 0.5 per cent of that of the industrialized countries. This may seem as modest amount. If there is still doubt as to whether these resources will be deployed where they are needed most, this is because agriculture cannot be viewed in isolation from national economies or from the world economy as a whole. The market economy is geared to purchasing power, not to basic needs, and the distribution of needs but, putting it simply, by the distribution of power.

Consequently there are two grim truths which we have to take as our starting point in reviewing the social prerequisites of a rise in food production.

1. "It is not surprising that food cannot be produced by people who cannot afford the costs. This gives the problem the illusory appearance of a food deficit. But malnutrition is not caused by this deficit. Malnutrition occurs because people cannot afford to eat. This problem cannot be solved by direct efforts to reduce population growth or increase the supply of food".(1)

2. Many people wonder whether the earth is capable of feeding a population which by the year 2000 will have passed the 6000 million mark. The answer is that it can already do so today, as witness table 11. Where protein production is concerned, the difference between 'needs' and net output today amounts to a surplus of 129 per cent. In terms of the kilos of calories produced, the surplus is 59 per cent.

Thus in many ways the food crisis is a crisis of distribution. This does not mean that we should refrain from efforts to increase total output - on the contrary. But it does mean that a rise in total output will not of itself have a direct bearing on the reduction of world starvation.

The distribution of power within and between the countries of the world does not only determine the distribution of output, it also decides what is to be produced.

(1) Professor Leonard Joy, "The Food Crisis and Third World Development - Impact for UK Policy", statement to a committee of the House of Commons.

Table 11. Annual world output of and need for agricultural products

	Protein, m. tons	Kcal 10^{12}
Net output	183	5857
Needs (3700m. people)	80	3700

Source: World Food Program and the Protein Problem (Copenhagen 1973)

The main idea underlying the theories of development which dominated the debate of the 1950s and 1960s was that, by participating in an international division of labour and in international trade, each country had specialized in the type of production which led to the greatest economic growth, which often meant the same thing as maximizing export revenues. "Growth first, distribution afterwards" has been the leitmotiv of development models, and growth has been imagined reaching everybody by a process of 'trickle down' through the community as far as the poorest members.

Today we know that this argument is untenable where a large proportion of the countries of the world are concerned, and especially where the developing countries are concerned. Internal power structures cause the economic surplus to be very unevenly distributed, and purchasing power does not reach those who are most in need of staple foods, for example. Their needs remain unsatisfied in the form of what is termed ineffective demand, while other groups in the community are able to use the country's export earnings to satisfy needs of a far less basic nature. Powerlessness and poverty therefore make for lower output of basic foodstuffs and have the effect of perverting the focus of production.

We ought therefore to refrain from making a sharp distinction between production and distribution. This realization has led to the formulation of the basic needs strategy mentioned previously, a strategy which implies the direct focussing of production on elementary needs. Briefly this may be said to imply that the content of production must be determined by unsatisfied wants. In other words, the "growth first, distribution afterwards" axiom is rejected in favour of "growth conducive to redistribution". There is no doubt that a strategy on these lines would boost the food output of the developing countries. An yet there are formidable obstacles to its implementation, simply because it raises the question as to who is to cultivate and what is to be cultivated, thereby touching on fundamental problems relating to the distribution of political and economic power in the developing countries.

In the majority of developing countries, the agrarian structure has not changed significantly since 1951, when the United Nations added land reform to its development programme. Instead, unsatisfactory tenancy systems have persisted and the capital intensive enclaves of multinational enterprises (units of production segregated from the rest of the country's economy) have developed new forms of production, mainly for export. At the same time as meat production is being developed in extensive areas of Latin America and Africa for export to the United States, Europe and Japan, starvation is becoming more widespread in the exporting countries.

Thus the agrarian structure of the developing countries and the structure of the international economy probably have a crucial bearing on the possibilities of exploiting the great potential for increased food production that is offered by physical and technical conditions.

To these social impediments must be added another grave factor of uncertainty, namely the possibility of environmental changes which could affect the production capacity. Desertification can result from acute droughts or perhaps even from long-term climatic change, but in most cases the devastation of cultivable land has resulted from human activity. Excessive grazing occurs for example, when newly bored wells make it possible to increase herd sizes. There are many places in Africa where deserts have been created in this way. Population pressure also leads to over-cultivation, unless systematic countermeasures are taken; a community cannot afford to leave any land fallow, bad harvests ensue and erosion follows in their wake. The wind can carry off a centimetre of soil every month. As mentioned above, the governments of many parts of Africa allow multinational farming enterprises to use large areas of their best farming land for export production. A policy of this kind multiplies the pressure of population on other areas thereby accelerating desertification. Recurrent droughts dramatically reinforce this process, giving it the appearance of an 'act of God'. Similarly, irrigation can cause the soil to lose fertility as a result of leeching, increased saline content or waterlogging. It is estimated that hundreds of thousands of hectares are eliminated every year because of inept strategies of irrigation.

Thus there are various ways in which human action can directly reduce the prospects of raising food output. On a more general plane, we have the uncertainty surrounding the effects on agricultural output of possible long-term climatic changes. Where agriculture is concerned, the main topic of discussion is a future decline in temperatures. There is a great deal of disagreement between scientists on this point, partly because the period internationally taken to represent 'normal climatic conditions', namely the years between 1930 and 1960, appears to have been one of the warmest thirty-year periods in the past thousand years. The questions now being discussed with reference to food production are, firstly, whether a return is now in progress to somewhat lower temperatures and, secondly what effects are likely to stem from the heavy fluctuations and frequent occurrences of extreme weather conditions (floods, droughts, etc.) associated with such a transition. A reduction of mean temperature - even a slight one - would have different effects on farming in different parts of the world. Variations in the weather, however, would have a general effect; farming in marginal areas (i.e. areas brought into cultivation to raise agricultural output and usually having natural potentialities inferior to those of areas earlier in cultivation) would be more susceptible. Thus uncertainty about the climate could increase at the same time as production becomes more sensitive to climatic change.

Finally we should mention a risk whose magnitude cannot be estimated today, namely the increasingly narrow genetic base of more and more of the world's crops. The advanced improvement of a small number of species can in the long-run increase the sensitivity of the species evolved and at the same time oust local variants which at some future date could prove invaluable as a starting point for the evolution of new and more resilient types. One means of retaining diversity in this respect is to increase the number of genetic banks in the world.

Can physical/technological possibilities, social/political impediments and ecological uncertainties be boiled down into a single assessment of the prospects concerning tomorrow's food production? Unfortunately they cannot. As has already been mentioned, the FAO compiled, in preparation for the World Food Conference of 1974, a forecast limited to the following ten years. Like all other forecasts, it had to be based on a number of very rough assumptions. The FAO forecast was based on the assumption of 'unchanged factors of change' in demand (as a function of population growth and rising incomes), as identified by analysing production during the 1960s. The outcome of this forecast is a balance between supply and demand at the global level but not at regional levels. It was assumed that the deficit (i.e. import requirement) of the developing countries would rise from 15-30 million tons to 85 million tons in 1985. The probability is, however, that the developing countries will be unable to afford purchases of food in such quantities, and today there is nothing to suggest that the industrialized countries would be prepared to furnish such quantities in the form of international assistance gifts. We will round off our discussion of future agricultural production with two observations concerning this forecast.

1. The forecast refers only to effective demand, i.e. demand generated by rising incomes. It does not include those people who remain too poor to 'demand' the food they need.

2. There is little likelihood of anything like 85 million tons being transferred to the developing countries. This equation, therefore, can only be solved <u>within</u> the developing countries themselves, which once again brings institutional conditions into the foreground. Even with very rapid and thoroughgoing reforms (which do not appear any more probable today than they did five or ten years ago), agricultural programmes would take time to put into effect. Unfortunately, therefore, it is a reasonable assumption that the dependence of the developing countries on imports of foodstuffs from the industrialized countries, above all from North America, will continue at least until 1985.

Whatever the development of dependence on imports, there is a great deal to suggest that starvation may increase during this period.

Forestry

Whereas agriculture produces one or more harvests every year, it takes at least fifteen or twenty years, sometimes up to a century, for trees to become ready for felling. The slowness of forest growth makes it easier in principle to judge the stock which will be available in the future and the felling that will be possible. On the other hand, a community may be tempted to extract more than it can replenish, thereby mortgaging its timber assets for many years to come. This at present is the situation in Sweden.

The amount which can be extracted from a given stock depends, of course, on the rate at which that stock grows. At a rough estimate we may assume that some 2.5 per cent of the world's stock of timber can be used every year without any reduction of new growth. It is difficult to calculate the size of the stock of timber, partly because there are so many different types of forest in which, moreover felling is subject to different conditions. The world's timber resources can be <u>roughly</u> estimated at <u>something in the region of</u> 300000m. cu. metres. Given an average possible estraction of 2.5 per cent, this would theoretically correspond to 7500m. cu. metres a year. On a very long-term basis and with more intensive methods of cultivation, it might be possible - again, theoretically speaking - to raise felling to between 10000 and 15000 million cu. metres. The global extraction rate today is 2500 million cu. metres. <u>At the very most</u>, therefore, the present-day extraction rate could conceivably be multiplied by between four and six times.

But the bounds of practical possibility for the coming twenty-five years are a good deal more constricted. Studies of future timber consumption indicate that it will double, reaching about 5000 million cu. m. by the year 2000. Purely technically speaking, there is no reason to doubt that the world's forests can produce double the amount they are producing today, so that there would be no need for a global timber shortage during the next twenty-five years. As always, however, global

Taking the world as a whole, firewood accounts for a considerable proportion of consumption of forest products. FAO statistics indicate that 80 per cent of the forest output of the developing countries is consumed as firewood, and in some countries it may be anything up to 90 or 95 per cent. (Picture: Yngve Baum/SIDA Picture Archives).

figures are misleading. This is because, in very schematic terms, we are dealing with two different types of countries, two different applications and two different kinds of forest or timber.

In Sweden, 95 per cent of the forest is used for industrial timber - timber products, pulp and paper, etc. - while 5 percent is used as firewood. As we saw in the preceding chapter, the situation in the developing countries is almost exactly the reverse, with 80 per cent being used as firewood. This means that about half of all the timber produced by the earth's forests is burned. Different conditions govern supply for these various purposes.

It is assumed that by the year 2000 the need for industrial timber will be at least twice what it is today. At present, roughly two-thirds of industrial timber is softwood, which mainly grows in the northern hemisphere. There is hardly any prospect of raising production in Europe. Instead the bulk of the increase will have to come from North America and the USSR. An increase in magnitude which would then be needed (2.8 per cent per annum) is probably on the boundaries of what is theoretically possible, and furthermore it would demand enormous investments in new roads and settlements. This in turn would raise timber prices, and so it is quite unlikely that the world's output of softwood can be doubled by the year 2000.

On the other hand hardwood production can definitely be doubled. But the replacement of softwood with hardwood will, among other things, require a development of the technology currently used in the manufacture of paper. The great unutilized reserves of hardwood are mainly to be found in a small number of heavily forested developing countries which will presumably be invited to take part in joint ventures aimed at the export of timber. But these countries may become less and less disposed to export unprocessed roundwood, stipulating instead that they be enabled to upgrade their raw materials themselves.

On the whole the present timber problem of the developing countries is an energy problem. It is not easy to estimate their future need of and effective demand for firewood. Many countries already have very severe shortages at local levels. Firewood differs from industrial timber in that local or regional shortages cannot be overcome by imports from distant areas because transport costs are prohibitive and the people who need the wood are poor. And yet firewood is an absolute necessity and is frequently one of the heaviest items of daily expenditure. Surveys from the capitals of Niger and Upper Volta have shown that an average working class family spends 25-30 per cent of its income on firewood. Since there is little that can be done to save the fuel used for cooking and heating, the need for firewood can be expected to rise steeply together with population growth, and the shortage of firewood can therefore be expected to do the same. One possible alternative to firewood is kerosene which has long been more expensive, though constantly rising wood prices have begun to close the gap despite the increases in petroleum prices.

A shortage of firewood has a number of serious consequences. It increases the pressure that is already being exerted on the vulnerable environment of arid and thinly wooded regions. People who lack fuel for their daily cooking cannot afford to think about long-term ecological balances. Secondly, shortage of wood causes dried manure to be used as fuel instead. In Southern Asia particularly, but also in the countries south of the Sahara, this practice is increasing rapidly, depriving the agricultural sector of an important nutrient. (The manure burned in India corresponds to one-third of the artificial fertilizer used in that country). This reduces food production and causes leeching of the soil, which in turn may give rise to long-term ecological damage.

Thirdly, shortage of firewood is liable to influence eating habits in parts of the world where there is already malnutrition, in such a way that people change over to food which can be cooked with less fuel. This can impair the composition of the staple diet. Fourthly, looking for firewood, gathering it and carrying it home over increasing distances in order to meet everyday requirements is occupying an increasing proportion of the energies of many poor families, energies that are really needed for tasks of a more long-term nature at both the local and national level.

Thus raw materials for household fuel is the really serious global problem in the forestry sector. Purely technically speaking, it seems as though a short-term solution of this problem could be achieved if part of the gigantic petroleum consumption of the industrialized countries could be diverted to the developing countries in the form of kerosene. But the same applies to this question as to the question of foodstuffs; the decisive factor is not needs but purchasing power and, behind purchasing power, the distribution of economic and political power.

For the immediate future, therefore, the problem of the world's forest assets is not a common or even a similar one between poor and affluent countries. If the earth's forests are nonetheless regarded collectively in a historical perspective, one finds that they have constituted a dwindling resource. A rough calculation indicates that human action through the various great civilizations represented by the river civilizations of the Near East, the Mediterranean civilizations, China, Europe, North and South America, etc. may have resulted in the elimination of no less than half the forests that originally covered the earth.

The forests of the developing countries are still diminishing by more than one per cent annually. There are many reasons for this. Perhaps the most important of them is forest clearance for farming purposes. Next comes the extraction of timber for fuel. Third comes the felling industrial timber (in Southeast Asia and Tropical Africa) for the industrialized nations. Thus at present the expansion of farming acreage constitutes the greatest threat to the forests of the developing countries. We observed in the preceding section that the FAO had set the target of a 20 per cent increase in the cultivated acreage of the developing countries between 1970 and 1985. What constitutes a promise in one dimension can thus amount to a menace in another.

The long-term solution is to try to replant the forests where they are destroyed, both to protect the environment and to guarantee a raw material base. Great progress towards this end has been made in China, where, starting in the 1950s, the gradual diminution of the country's forests has been reversed for the first time in 5000 years.[(1)] There is still a shortage of wood, and only a small proportion of the trees planted survive, but the trend has been broken. Elsewhere in the world, reaforestation programmes have usually been less successful. It is immensely difficult to protect growing stands from the depredations of destitute people in search of cultivable land and fuel for their cooking fires. Here again we find that the possibilities of utilizing - and also of preserving resources are conditional by economic and political factors.

Fisheries

At present between 60 and 70 million tons of fish are extracted from the oceans. Potential catches of traditionally caught species are thought to exceed 100 million tons; landings could thus perhaps be doubled. This holds good despite the well-known fact that many species are over-fished in the North Sea, the Atlantic, the

(1) Also Sweden is probably one of the rare examples in that the forests have doubled over the last 70-100 years

North Pacific and elsewhere. Other species in the same areas and in the Indian Ocean are far from exhausted. At least, so far fishing crises only apply to particularly attractive species and to certain regions.

If we include other species besides the traditional ones, e.g.krill and squid, the present world catch could be multiplied several times over, possibly rising to something like 300 or 400 million tons. Of course, figures like these are highly uncertain, but they still have something important to tell us: whereas world catches multiplied eighteen times between 1900 and 1970, no such increase can be expected in future. As far as we can see today, a quintupling of catches marks the ceiling, and perhaps a doubling of catches marks the limit of what is practically possible where traditional species are concerned. No allowance has been made here for the increments which could result from completely new forms of marine cultivation, partly because costs so far do not suggest that fish farming could change the picture. As was pointed out in the preceding chapter, demand is considered the most powerful factor restraining the growth of fishing for human consumption. At present roughly one-third of the fish caught in the world is ground down as industrial fish into oil and fish meal.

Minerals

"We will run out of silver in eleven years' time!" - "Nonsense, there's enough to last for 65 million years!"(1)

In the preceding chapter a summary was given of various forecasts of demand for minerals. It is worth emphasizing that the more consistently thought-out forecasts always refer to demand. In current forecasting, the entities on the supply side - production, reserves and resources - are regarded as variables which have to be adjusted to demand. Thus, as we saw in Chapter 2, the concepts of reserves and resources have not been designed to provide a foundation for independent forecasts of supply. If they are nonetheless applied to such a purpose, the outcome can vary as widely as in the above quotation.

Seen in a twenty-five year perspective, estimates of reserves are the main topic of interest. Table 12 illustrates the magnitude of known reserves today in relation to anticipated demand.

Reserves of the great majority of minerals amount to no more than one or two times the total consumption anticipated for the period ending in the year 2000. Is not this a very serious state of affairs? At all events, it is not as serious as the figures suggest. The main reason why reserves are not greater is that prospecting is expensive and the majority of the companies which carry out prospecting are privately owned and therefore have to employ strict profitability criteria. To a company with the customary requirements concerning the return on capital invested, it would be squandering resources to spend money on expensive prospecting for deposits that will not be drawn on until after the twenty or thirty years marking the normal horizon for long-term planning. In a word, reserves lasting for fifty or a hundred years are a luxury which hardly any company feels it can afford. A newspaper headline not long ago summed up this state of affairs rather drastically, though succinctly, in the words: "How long will reserves last? Thirty years at a time".

If reserves, expressed as a number of years' consumption, have changed at all in recent decades, they have tended to increase rather than diminish. This does not necessarily imply that they will do so in the future.

(1) Those are examples of rough interpretations of data that emanate from the US Bureau of Mines (1972) one one hand and from Annales des Mines for 1971 on the other

Two kinds of state fishery. Above, a long well-filled trawl being winched on board the factory ship Jung Garde, 10,000 tons, which forms part of the large East German fleet, fishing on an industrial scale off the coasts of Canada and Greenland. Below, state carp farming on the shoreline in Orissa, Eastern India. (Pictures: Zentralbild, DDR, and S. Berg/SIDA Picture Archives).

Table 12. Demand between 1974 and the year 2000, and reserves of certain common minerals.

Minerals	Units in metric tons	World demand 1974-2000	World reserves in 1974	Reserves in relation to cumulative demand 1974-2000
Aluminium	Million	873	3483	4,0
Cobolt	"	1,16	2,45	2,1
Copper	"	320	408	1,3
Chromium	"	92	523	5,7
Gold	Thousand	32,8	41,1	1,3
Iron ore	Million	20000	91000	4,5
Lead	"	125	150	1,2
Manganese	"	370	1826	4,9
Molybdenum	"	4,08	5,90	1,4
Nickel	"	26	54	2,1
Phosphate	"	6175	16065	2,6
Silver	Thousand	420	187	0,4
Sulphur	Million	2036	2040	1,0
Tin	"	7,65	10,32	1,3
Tungsten	"	1,48	1,78	1,2
Zinc	"	217	236	1,1

Source: US Bureau of Mines/United States Department of the Interior, "Mineral Facts and Problems", 1975 Edition. (For definitions and methods of calculation, see report 311 from the RoR study, 1977).

On the contrary, it may reflect a belief on the part of mining enterprises that there will be a growing shortage in future and that this justifies intensified prospecting in order to pre-empt those deposits which are still relatively cheap today. But this is only one interpretation. There may also be other intentions behind the figures published by companies concerning reserves. As a general rule, however, figures for reserves are a gross underestimate of the deposits which actually exist and can be remuneratively extracted. Furthermore, the relative cost of producing a given quantity of minerals has so far declined in relation to the cost of other products; on this point, see pp. 70, 86.

There are many reasons for critically examining reserve figures even as indicators of what they are supposed to reveal, namely deposits which have been discovered and are worth exploiting at the present day level of prices. Mineral extraction is a field involving a small number of actors (companies and countries), and those actors may often have economic or political reasons for overestimating or under-estimating the size of reserves. It is impossible to make allowance for this in total figures of the kind with which we are dealing, but we still believe that an overall assessment indicates that a continuation of the present growth rate of consumption would cause the prices of certain metals to go up. These rising prices would signify that the remaining natural concentrations of the metals in question were far inferior in quality to those which had already been used. Over a period of twenty-five years, this is perhaps most likely to happen to tin and silver. Prices of copper and tungsten may also go up, and in a more long-term perspective (50-75 years) tendencies of this kind are also likely to affect both other base metals and precious metals.

It is not actually certain, however, that prices will go up. Whether they do so or not will depend on the extent to which the raw material in question can be replaced by others, Mercury is one metal whose price shows a declining trend even though reserves are quite limited in relation to annual consumption. This is at least partly because the use of mercury is being steadily reduced for environmental reasons.

Our general conclusion is clear, however. As far as the great majority of non-energy minerals are concerned, there exist today known and remunerative reserves which substantially exceed the total demand anticipated for the period ending in the year 2000. Nor is there any reason to suppose that they will be suddenly exhausted during the period immediately following the year 2000. It is only in physical terms, however, that this conclusion can be stated quite unambiguously. It does not by any means imply that supplies of minerals will be free from problems either globally or in Sweden, even during the next few decades. The following are the most important of the factors complicating the physical perspective.

Growing Energy and Pollution Problems

There is no doubt that energy will constitute the most important problem of resources during the coming twenty-five years. Practically all observers agree that a physical shortage of oil will occur within a few decades. Rising energy prices are probably to be expected. At the same time, the change to increasingly poor orebodies will demand the handling of progressively larger amounts of minerals and will thus entail rising energy consumption. Energy use can probably be made more efficient up to a point, but beyond this point the rising level of energy consumption per ton of metal will have the impact shown in figure 27.

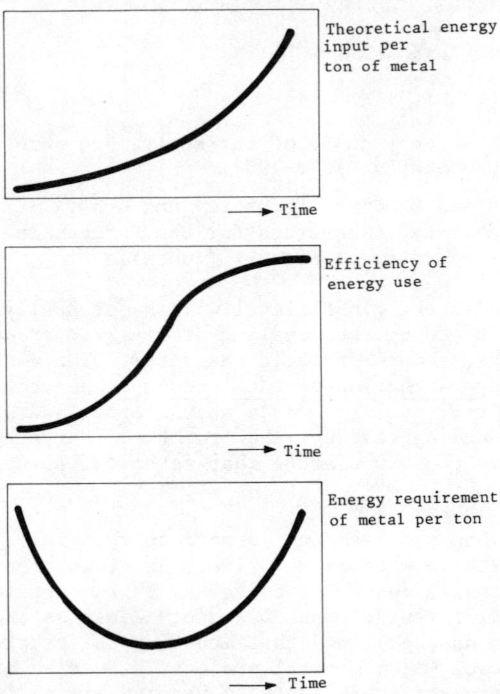

Figure 27. Theoretical energy consumption in the production of minerals.

Source: Lars Kristoferson in "Save or Squander" (ed. Sten Schager), p.116, Ingenjorsforlaget 1977. (In Swedish)

These developments are unlikely to have any noticeable impact within the next twenty-five years. Probably they will take the form of a more long-term tendency which may be hard to discern and, in a more short-term perspective, may even be eclipsed by compensatory factors. It is obvious that the total potential for development is a fundamental determinant of the extent to which rising energy requirements and energy costs will affect mineral prices. Hitherto, at least as far as has been recorded up to the late 1950s,costs per unit of output in extractive industry have declined, as can be seen (up to 1957) from figure 28.

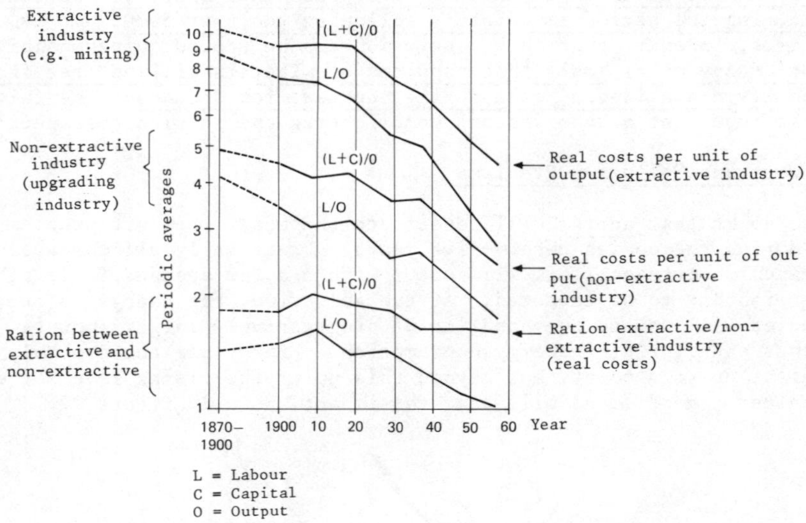

Figure 28. Development of unit costs of extractive and non-extractive industry in the United States, 1870-1957.

Source: H J Barnett and C Morse, "Scarcity and Growth - The Economics of Natural Resource Availability", Resources for the Future, The Johns Hopkins Press, 1963. The scale refers to relative figures (index).

Whereas energy problems at the global level will be gradually accentuated, the environmental problems posed by the handling of minerals are of a more immediate and, in certain respects, a more dramatic character. The external environment is affected by large-scale destruction of the landscape, as occurs in open-cast mining of low grade ores resulting in large waste heaps, etc., and also by the diffusion of heavy metals from leeching residues and from the first stages of processing. In both cases, there are those who argue that technical solutions can be found to these problems.

No convincing evidence has yet been put forward to show that large-scale mineral extraction can be undertaken without relatively heavy environmental costs. All we can say with a reasonable degree of certainty is that these costs tend to be valued differently by different groups in society (just as their impact naturally varies from one group to another), and that mounting political conflicts can be expected on the subject of major mineral projects. Swedish examples have included Ranstad, the Kambrium project in Skane (both uranium mining in fertile agricultural areas) and the possible extraction of minerals in the Scandinavian mountain range. Conflicts on similar projects can be foreseen, insofar as they have not already materialized, in Western Europe and the United States and may affect both the location and the expansion rates of large new mineral extraction facilities.

Open-cast mines and their associated facilities often take up tens of square kilometres, but the environmental effects extend over much greater areas. Increasing mechanization during the past century has prevented costs per ton from rising in relation to other goods, despite the extraction of more low grade ores. The picture comes from Kounradskiy in Central Kazahkstan. (Picture I. Budnievitch, Fotochronika, TASS). The mining operations of earlier ages involved a heavier input of human labour, as can be seen from the next picture which comes from the second German edition of the De Re Metallica of Georgius Agricola, published in Frankfurt am Main in 1580: (p.72)

In addition to the problems of the external environment, we have those of the occupational environment. Metal mines in Sweden lose far more working days on account of accidents than any other sector of employment (over 4.5 times the average for all sectors), and the same probably holds good in other countries as well. In the United States, for example, recruitment of face workers is regarded as one of the great bottlenecks in efforts to boost coal output. Stipulations concerning both the external and the internal environment may be watered down in times of economic difficulty and uncertain employment and are thus a function of the economic situation. It is a well known fact that few governments in the developing countries feel able to afford the same environmental standards as are practised in the industrialized countries. But the environmental problems of the developing countries are mostly of a different nature, viz malnutrition, polluted drinking water, overcrowding and unhygienic conditions. It has been discussed whether this difference in attitudes will result in the near future in a larger proportion of mineral production and, above all, a larger proportion of the polluting processing industries being established in the developing countries. Willingness to accept environmental damage is undoubtedly a comparative advantage in a free world market, seen from the point of view of the enterprise, and it is therefore not unreasonable to anticipate developments in this direction.

We may sum up by saying that rising energy costs and rising environmental costs both impose restrictions on the long-term continuation of the rise in consumption of metals. These costs may in fact come to influence the rate of expansion and the location of new mineral extraction facilities within the next twenty-five years. This especially applies if rising costs really materialize as a consequence of more people valuing the protection of the environment higher than before.

A
D
C
B
C
L
G
I
K
H

Increasing Difficulties in Mobilizing Reserves

The international discussion concerning mineral supplies for the next few decades has not concentrated to any great extent on the question marks surrounding the topics of energy and the environment. Instead the main preoccupation has been with the capacity of the Third World economies to mobilize and organize their resources in such a way that the quantities demanded will be delivered without any serious disturbances and, secondly, the possible effects of political and military developments in the Third World and of efforts to form producer cartels on a global scale.

There are many different ways of looking at the difficulties involved in inadequate investments in a medium-term perspective. The Korean boom and the steep rise in prices during the early 1950s were followed by an expansion of the mining industry, which created a relative abundant supply of minerals during the 1960s. This lasted until the 'raw materials crisis' of 1972-74. The industrialized countries are now uneasy because, so far at least, the price rises of the present decade have not generated a new wave of investment resembling that which occurred in the 1950s. This is partly because, after 1974, prices fell once more to a relatively low level a level which was high enough for most older mining facilities to pay their way but not high enough to cover the substantially greater expenses which most new mining projects could be expected to involve. The older mines are often better situated, with richer ores and better infrastructures, all of which helps them to operate profitably even at lower prices. Where major projects are concerned, a long period of time separates the decision to go ahead from the delivery of the first consignment or ore. And the further ahead in time prices have to be calculated, the more uncertain they become. There are any number of arguments and counter-arguments on this subject of the difference between today's and tomorrow's mining costs. All that we can say with any certainty is that the problem does not by any means apply to all minerals. Copper is the most assiduously quoted example. One conceivable pattern of development is that increased output will be obtained principally by expanding the capacity which already exists, i.e. working existing facilities for all they are worth, so that a shortage will ensue when these mines begin to be worked out and new mines have not yet been opened up for production (which can take up to five years). Problems of this kind could result in a jerky flow of copper supplies accompanied by severe recurrent 'crises' and price freezes.

Another line of arguement concerning under-investment in the minerals sector focusses on the geographical distribution of investments. Widespread nationalizations of foreign-owned mines in the Third World, coupled with the greater magnitude of new mining projects and the longer time which they now take to put into effect, has led people to conclude that the risks involved by investing in mineral projects in the developing countries have become greater. Companies therefore prefer to exploit low-grade ores in areas which may have inferior natural potentialities but are located in more 'reliable' countries such as Canada and Australia. In 1975 an American survey of the investment plans of mining enterprises for the next few years showed that nine-tenths of all foreign U.S. investments were going to Canada and Australia, and that the investments planned for the Third World countries were negligible. Here again, opinions vary as to whether or not this marks the beginning of a new long-term trend. If it does, the effect will be for low-grade ores to be exploited and superior ones to be left unexploited, which will mean rising mineral prices. Another argument is that the risk assessments made by business enterprise are an expression of the tensions in the world, and that these tensions lead to an increasingly non-optimal extraction of mineral resources. This is hardly a surprising turn of events. Instead it is the logical corollary of the patently uneven distribution of the consumption of minerals (like the consumption of other resources); the tensions between the industrialized and the developing countries are principally brought about by this unevenness.

A third difficulty attending the mobilization of resources for the supply of mineral may be caused by what we may term diminishing efficacy. The further below ground and the more inaccessible the areas where prospecting has to be undertaken, the less attractive the ores become and the greater the amounts of stone and waste which have to be crushed, handled and transported, and the greater will be the amount of machinery, crushing plant, transport and other inputs required for every unit of finished metal that is produced. So far, technological progress has been so rapid that the unit cost of minerals has shown a declining trend: cf. figure 28. But a ceiling for efficiency gains may be reached in the mineral sector just as in energy production. This would mean a rise in the total expenditure incurred by the community for the procurement of a given quantity of new materials. The more things - e.g. environmental costs - one includes in the category of material costs, the greater will be the increase in the cost of materials to society. In monetary terms this is manifested by a shortage of capital. The amounts which have to be invested to produce a given quantity of raw materials grow larger and larger. This would imply a species of diminishing total returns within the raw material sectors, but it is uncertain whether events are actually heading in this direction. The other side of the same coin concerns capital yield as such in the mineral sector. Lower than average profit levels, diminishing possibilities of internally financing for new projects, uncertainty as to when the higher costs could be recouped via higher prices, etc., might all lead mineral enterprises to diversify their activities and invest their surpluses in other sectors instead of mineral extraction. This, at least, is contended by several American corporations.

When the large mineral enterprises discuss the low level of profits in the sector, this is not unconnected with their wishes in relation to the state and, ultimately, in relation to consumers. These wishes may concern tax reductions, prospecting grants, boarder protection (tariff walls, etc.) - in short, measures which can boost the return on capital invested. But self-interest thus shown by business enterprise does not justify a general rejection of its arguments. An increase in the machinery, working hours and energy required for every ton of minerals extracted will increase the cost of raw materials to the community as a whole. But governments depend on mineral enterprises for the information they need in order to size up the arguments propounded. As we have already noted on the subject of reserves, impartial and neutral information is probably a rare commodity in these contexts.

Interstate Cartels

Following the unique success achieved by the petroleum-exporting countries through the OPEC cartel - which practically quadrupled the price of oil overnight - the question of interstate cartels seemed for a time to be the most important issue connected with raw materials. The industrialized countries were afraid that similar cartels might be set up for a number of other raw materials. The developing countries, for their part, hoped that associations of producers would lead to heavy increases in their export revenues. An important part of the international debate on raw materials between 1974 and 1976 centred round standpoints concerning co-operation between producer countries.

A cartel is an association formed by the producers of a particular commodity for the purpose of raising their profits. Often the intention is for this to be achieved by limiting output (and supply). OPEC is an exception to this rule because it has not set any limits to output; its decisions have only been concerned with prices. In order for revenues to rise, demand must be so insensitive to price increases that the advantages of the price increase will not be offset by a contraction of demand. Demand for many raw materials is generally taken to be relatively insensitive in the short and medium term. This applied, at all events, to the prices of raw materials until the 1970s, which were low in themselves and, furthermore, underwent long periods of decline. Demand will not necessarily remain as insensitive if there is a rise in the relative prices of raw materials, i.e. if raw material costs come to account for an increasing proportion of the price of the finished product.

In the long run this will increase the possibilities of replacing one raw material with another. For example, a steep and lasting rise in the price of copper will probably accelerate a change to aluminium. Natural rubber has to a great extent been ousted by synthetic rubber, partly because the price of the latter has been relatively stable.

A number of other conditions have to be satisfied if a cartel is to be effective for any considerable period of time. The members of a cartel should account for a considerable proportion of world output and/or of world exports. Otherwise it is quite probable that non-members will simply step up their production, causing the members to lose ground in the market.

But even if the members control a dominant share of world trade, the price rise achieved by the cartel may relatively soon lead to increases in the amount supplied by other - possibly new - producers. Bauxite, the raw material from which aluminium is extracted, is one possible example. If IBA, the present producers' organization, were to succeed, say, in doubling the price, it is believed that it would become viable to extract aluminium from other raw materials, including certain clays which are found in large quantities in the industrialized countries.

Cohesion between the member states is important. Once the price has been pushed up by means of output restrictions, each individual country will then stand to gain by raising its production. The very risk of the cartel breaking down as a result of individual members producing more than has been agreed on is a powerful incentive to each country to do exactly this; the country which keeps its production down longest is the country which will lose most if the cartel collapses. Thus the countries determining the policy of supply - the producer countries - must have a powerful sense of common interest vis-a-vis the consumers in order for effective co-operation to be established. A large producer in Canada or Australia, for example, may feel that it is more in line with his country's interests to maintain good economic relations with the consumer countries. (There are many firms in the raw materials sector to whom this thesis applies because they are owned by interests in the industrialized countries).

The above list of difficulties involved in holding a cartel together for any considerable period reflects arguments presented in the majority of survey reports concerning cartel opportunities or hazards. The conclusion drawn in all of these reports is that effective and longlasting cartels are unlikely for other raw materials than oil. The possibilities of producer co-operation are perhaps slightly greater in the case of copper, bauxite, chromium and tungsten than for other raw materials. Bananas and tropical woods could also be included in this category.

But perhaps the idea of new cartels has been too easily dismissed. Firstly, one finds that firms in the raw materials sector often seem to have succeeded to establishing quite advanced co-operation. The Seven Sisters, i.e. the big oil companies, existed long before OPEC. Their informal but nonetheless effective co-operation lives on today, despite OPEC, as witness the curious uniformity of their prices. There is co-operation of different kinds between the big aluminium companies, between pulp and paper manufactures, between steel manufacturers, between producers of gold and diamonds, and so on. There is not enough room in this final report for an extensive discussion of the role of the transnational enterprises in raw material markets, but clearly they play an important part, not only in estimates of resources - a topic which we have already considered - but also in the choice of technology, in trade flows and in co-operation between industrialized and developing countries.

In view of the way in which the transnational enterprises dominate markets, greater interstate co-operation is the only chance the developing countries have of mobilizing their resources and exercising 'co-determination'. In this process the state tends to acquire more and more influence in the raw materials sector.

Although there are great difficulties involved in keeping an interstate cartel intact for any great period, more shortlived cartels may very well appear. To governments which are short of foreign exchange and which give highest priority to an increase in production the short-term gains accruing from a cartel may be worth even a long-term loss.

So far we have concentrated most of our attention of the most difficult of all the various possible forms of producer co-operation, namely price cartels. There is in fact a complete spectrum of co-operation, ranging from a straightforward interchange of information concerning, for example, such matters as investment plans, to the control of prices and production.

All in all, cartels appear unlikely to exert any decisive influence on the pattern of world demand for raw materials over the next twenty-five years. Producer co-operation of different kinds may rock the markets for brief periods where individual raw materials are concerned, and episodes of this kind can undoubtedly cause the consuming countries trouble in the short term.

Political Events

Production of certain minerals (e.g. tungsten, chromium, manganese, vanadium, cobalt and nickel) is concentrated in a small number of countries, and as a rule the number of countries accounting for the dominant proportion of the world's exports is smaller still. It is therefore natural that regional conflicts involving major exporting countries should attract attention in the context of medium and long-term supply. Chromium, platinum group metals, diamonds from Southern Africa, copper from the African copper belt, tin and rubber from Southeast Asia - not to mention Middle Eastern oil, of course - are a few instances where the exporting countries provide such a large share of the global supply that political events within or between the countries in each region can have a serious short-term impact on supplies to the industrialized countries. It is perhaps typical that a joint British and North American committee should have pointed to the possibility of a shortage of chromium resulting from the political situation in South Africa. Realizations of this kind have caused the great powers to take a closer interest in these areas and have probably increased their tendency to try to influence local and regional policies in each area. Prolonged conflicts involving countries of this kind will cause peacetime crises (i.e. disruptions of supply) elsewhere in the world. Because markets are sensitive and the tendency to over-react and speculate is never far away, the actual size of the conflict may be dwarfed by the effects it has on the raw materials sector.

The politicial developments associated with peacetime crises are most often characterized by armed conflicts of one kind or another. These conflicts often accompany political changes in the developing countries. In the long run the direction of political change may lead to a reduction in the supply of raw materials, especially renewable ones like oils and fats, oil cake, fish meal, meat and certain plantation crops. This would probably be the effect of structural reforms which focussed the economies of the developing countries on the needs of their own populations instead of, as at present, on production primarily for the international market.

Thus internal politics in the Third World do a great deal to determine what is not produced as well as what is. In the interim study entitled "Renewable Raw Materials" an analysis was made of the reasons why the power structure can inhibit production. For example, because of the agrarian structure in Latin America, large areas of land are left uncultivated or used for the production of meat for export, at the same time as the local population is unable to get the staple foods it needs because it lacks purchasing power. Agricultural modernization in the developing countries often proceeds along the same lines of intensive capital and energy input as the most modern agriculture in the industrialized countries. No doubt this makes it possible to raise production, but the additional output is destined for exports or for markets within the producing countries. Because purchasing power is not created within the country in parallel with the rise in output - on the contrary, more and more people are ejected from the farming sector as a result of ongoing mechanization and large-scale enterprise - and because demand in the world market is rather rigid, many developing countries run the risk of developing an agricultural sector which runs at a loss at the same time as distress within their own borders continues to grow. In 1976, India sold two million tons of wheat in the world market. This was not due to a surplus at home, it was merely due to the silos being overfilled.

In a more long-term perspective, the governing strata in the majority of developing countries may conceivably lose the community of interest - based on continuous export revenues and arms shipments - with enterprises in the industrialized countries. If this were to take place, we can outline the contours of a more overtly north-south conflict. A conflict of this kind, if it occurred, would be characterized by genuine exploitation of the trade in raw materials by the developing countries in a common, co-ordinated struggle to wrest from the industrialized countries higher prices for raw materials, greater access to markets for finished products and various financial and other concessions. If exports of renewable raw materials (primarily foodstuffs) and finished goods from the industrialized countries were to be heavily reduced as a retaliation, exports of non-renewable raw materials, i.e. minerals, could be the principle lever of power to be manipulated by the developing countries. Of course, this seems a remote contingency today, and we have therefore assumed the risk or possibility of effective interstate cartels to be fairly slight for the coming twenty-five years. But our argument is designed to underline the fundamental role which political changes can play in the raw materials sector, as well as the great difficulties involved in predicting such changes.

Summary

1. Our discussion can be summarized as follows with regard to physical restrictions.

A. There are no physical limits to restrict a rise in the supply of raw materials for the period ending in the year 2000, to about twice its present level.

B. On the other hand we can already discern the theoretical limits to a sustainable production of renewable raw materials, particularly fish and timber. The theoretical maximum production are something in the region of five or six times the present-day level or roughly three times greater than the level predicted for the year 2000. But the theorectical nature of these limits implies that in practice they will probably be unattainable, partly for reasons of expense. Furthermore, the yield from the world's forests, the area of which has perhaps been halved as a result of human activity, will depend on the progress of the conflict between farming and forestry.

It is harder to determine the theoretical limits for foodstuffs, but they are probably a little further away than the fivefold or sixfold increase indicated for fisheries and forestry. Here too, however, the theorectical limits are less interesting than the practical ones.

Even if it is the practical limits that are most interesting, the theoretical limits in themselves enable us to draw an important conclusion concerning renewable raw materials the increase in production cannot be maintained very long after the turn of the century. Since the dawn of the Industrial Revolution in Europe, the world's extraction of raw materials has multiplied several times over. Table 13 is a summary of historical figures, rough estimates of developments up to the year 2000 (based on 'conventional wisdom') and speculations concerning various theoretical limits. It should, however, be emphasized yet again that these figures are so uncertain that they can only be used as rough gauges of relative magnitudes.

Table 13. Population and annual production of raw materials at different times

	1870/80	1975	2000	Theoretical limit	Growth factor: 1975 from 1870/80	Growth factor: Theoretical limit compared with 1975
Population (1000 millions)	1,3	4	6	16-40 ?	3	4-10
Grain products (wheat, rice, foddergrain) (1000m tons)	0,25-0,3	1,4	2,7(?)	>10 ?	5	10-30
Forest (1000m.cu.metres)	1,5 -1,7	2,5	5	10-15	1,5	4-6
Fisheries (millions of tons)	<4	70	100	300-400	>18	5

Sources: FAO Commodity Review & Outlook 1975/76 and W. Vigener, "Die Weltprodulction wichtiger Waren" Berlin, 1970.

C. The non-renewable raw materials, paradoxically enough, are not subject to the same limits as the renewable ones, always provided that a solution is found to the energy problems of the future. The latter, as we have already noted, is a very important condition whose fulfilment is by no means to be taken for granted. The difference between renewable and non-renewable raw materials is an important one. In the case of non-renewable raw materials, there is no risk of the very basis of production, i.e. the natural foundation, deteriorating as a result of erosion or desertification, for example. In addition, the prospects of recycling are much greater for metals than, say, for timber and foodstuffs. Finally, the basis for substitution and the technological development of completely new and abundantly available materials is far greater where the non-renewable raw materials are concerned. It is therefore difficult to lay down any theoretical limits to the utilization of mineral-based products as a group. On the other hand, it is probable that individual metals will become exhausted in the sense of becoming so expensive that they cease to be in demand and therefore disappear from the economy, at least in a long-term perspective. Within the twenty-five-year perspective which we have been discussing, an incipient shortage may be manifested by gradual rises in the relative prices of individual minerals and by investments in research aimed at developing substitutes.

2. The following observations can be made concerning the political limits.

Certain physical limits, as was stated above, are already discernible today, although we do not believe that they will have any substantial impact this side of the year 2000. This is only a small part of the truth, however, so long as we omit to ask: impact of whom? This brings us onto the subject of political limits. Global figures, of course, tell us nothing about the inequality of distribution. "The limits to growth" have long since made themselves felt among the fifty per cent of the human race constituted by the poor of the developing countries. But these limits are political and social, not physical.(1) Even if the problems connected with energy and the environment can be solved, or at least deferred until after the year 2000, social problems and restrictions will influence both the total supply and the distribution of raw materials. Uneven distribution of political and economic power, oppression and growing gaps within and between countries generate contradictions which stand in the way of an extraction of raw materials that would be 'optimum' in the eyes of the affluent countries. The concentration of information and technology in the hands of a few transnational enterprises in the raw material sectors reduces the ability of governments to appraise the situation correctly and to implement a deliberate policy. Cartels and political conflicts may aggravate the sensitivity and turbulence of raw material markets, many of which are already characterized by violent and extreme fluctuations. These and other factors may lead to special supply measures being taken, but they make little impression on our basic conclusion: access to raw materials is unlikely to be a serious problem in the affluent countries for the remainder of this century. Insofar as serious economic problems are to be expected, they will not arise out of a shortage of raw materials. In many poor countries, on the other hand, shortages of energy, artificial fertilizer, wood and - probably - foodstuffs will constitute the worst problems of the next few decades.

Seen in a twenty-five year perspective the problems connected with raw materials are not quite so uniform as would seem to be suggested by a global approach and by mottoes like 'Only One World' and "in the same boat". As long as the total distribution of economic and political strength in the world remains extremely uneven, there will continue to be a difference between the raw material problems of affluent countries and poor ones.

(1) Sec e.g. F M Lappe and J Collins: "Food First - Beyond the Myth of Scarcity", Houghton Mifflin Co., Boston, 1977.

5 Resources in a Long-term Perspective

Introduction

In this chapter we shall be considering questions of resources and raw materials in a slightly more long-term perspective than previously. We have already seen how these questions are affected by a host of different factors, so that it is difficult to make any hard and fast predictions even twenty or twenty-five years ahead.

This uncertainty naturally becomes even greater if we try to look further ahead. The question is whether we can even claim to be making qualified guesses, and it should therefore be emphasized that the following arguments are much more conjectural than those propounded in preceding sections of this book. If, however, one were to desist from treating the long-term perspective on account of the great uncertainties involved, this could be justifiably said to result in a warped picture of the entire resource situation. The point is that assessing a certain factor to have its major impact beyond the twenty-five year period we have so far been dealing with does not automatically imply that this factor can be disregarded for another twenty or thirty years. In many cases, the adjustment period needed by society may be so long that a change will already have to be started in the 'near future' in order for satisfactory development alternatives to become available.

Time perspectives

What is a short time and what is a long time? Company investment and planning horizons seldom exceed fifteen or twenty years, but there are some connections in which several decades need not amount to a very long period. Many capital goods, for example, have an average service life in the region of twenty or thirty years. Shipping often comes in this category. Buildings and installations of different kinds often have longer service lives still. Forestry also demands distant planning horizons - perhaps as much as a hundred years, owing to the length of time it takes for a tree to become fully grown.

It has often taken far more than a couple of decades for an invention to be put to commercial use. The mechanical cotton harvester, for example, was not commercially exploited until over fifty years after it had been invented. The same thing happened to the gyro compass. It can take up to half a century for one raw material to be more or less completely substituted by another, as in the cases of natural and synthetic rubber and natural and synthetic fibres.[1]

It is more difficult, not to say impossible, to generalize on the strength of historical experience concerning the time needed to accomplish political, social and economic changes. History provides us with examples of political events which have rapidly transformed society - revolutions, for example - and also of gradual processes of economic and social change that have lasted for long periods. The Industrial Revolution, wherever and whenever it is considered to have occurred, was not a sudden event but a fairly long-drawn-out process. The same goes for its social consequences.

With due reservation for the great uncertainty surrounding long-term speculation, we shall now endeavour to look between fifty and hundred years ahead.

The 'model of doom'

Pronouncements concerning the future of the world commonly take as their starting point an observation of the growth of world population, and an assumption that the earth resembles a loaf of bread which is gradually being eaten up. This kind of model of doom was constructed in the first report to the Club of Rome, "Limits to Growth", and it had a very powerful pedagogical impetus.

The picture of the future reproduced in figure 29 was presented on the strength of historical time series for population growth, consumption of non-renewable natural resources, food consumption per capita and pollution of the environment. The writers presumed that the physical, economic and social relations applying historically to the world's development would continue to apply in the future.

The message was straightforward enough. We are living on a finite planet with a given 'stock' of non-renewable natural resources. There are also limits to the production of such renewable resources as agricultural products, timber and fish. Continued exponential population growth and economic expansion will thus lead to the exhaustion of the resource base. This will occur some time fairly early in the next century.

A picture of this kind has an undeniable instructional value as a means of showing what can happen subject to certain conditions and on the assumption that nothing is done to change the situation. But it is unlikely that this model conveys any information concerning what is going to happen. In fact the Club of Rome did not intend to give the impression of an inevitable future disaster. However, an insufficient analysis of compensatory factors made it easy for more or less careless readers to interpret the course of development thus outlined as a forecast, i.e. something more than a warning and a call to action.

In the following discussion of the 'model of doom' we shall be referring primarily,

(1) According to F J Jageler "Naturliche Rohstoffe im Substitutions-weltbewerb - Das Beispiel des Naturkautschukes", Hambury 1976, the global rate of natural rubber versus all other forms of rubber consumption in the world was in the mid-70ies roughly 1/3. In 1956/57 it was still roughly 55%. According J C Fisher and R M Pry "A simple substitution Model of Technological Change", Technological Forecasting and Social Change 3, 1971, the year 1956 was the year when synthetic rubber had taken over 50% of the production of natural rubber (counted in weight). The substitution period over which the new product goes from 10% to 90% of the market was estimated to be 58 years. In the case of synthetic fibres versus natural fibres the corresponding period was considered to be 58 years as well. The midpoint of the substitution process (where 50% of the market was taken by the new product) was however shifted in time relative to the rubber case to the year 1969.

not to the special models constructed on behalf of the Club of Rome[1] but to a more general mode of reasoning which we shall proceed to examine at closer quarters, (however starting the discussion with the more concrete example of the standard run of world 3 of the 'Limits of Growth' models). The most interesting aspect of a 'model of doom' of this kind is not the conclusions arrived at, but the assumptions and attitudes which the model expresses. The plausibility of the model of doom depends on the realism of the assumptions it incorporates, i.e. the faithfulness with which it portrays reality. This question may apply to the particulars conveyed by the model concerning the existing volumes of different natural resources and the quantities which will be available (energy included). It may also apply to the ecological, economic or organizational restrictions which may limit the availability of resources and the potential achievements of environmental engineering. Last but not least, it may apply to an assessment of the styles according to which people are willing and able to frame their lives. We shall dwell on some of the concepts, relations and assumptions employed, and we shall attempt to show that a traditional model of doom of the kind outlined above is too simple and too mechanical. For the sake of lucidity, the elements of models of doom - population, foodstuffs, natural resources and pollution - will be discussed individually and in that order.

Population Trends

As we have already seen, population growth carries a tremendous amount of weight in the model of doom. This is hardly surprising considering that the world's population is now doubling at intervals of thirty-five years. But population growth rates today do not convey any direct information concerning the resource situation of the future. A premature linkage between them can lead to fallacies of two kinds.

Firstly, the view is sometimes taken that the problem of an increasing scarcity of resources will be solved if only the growth of population can be restrained. While conceding the necessity of eventually inducing present-day population growth to level out, we do not believe that this will be sufficient to safeguard against the occurrence of resource problems in our more long-term perspective.

The second type of fallacy lies in the assumption that population growth will automatically continue as it has done so far, until the population of the world is too large for all its members to be assured of even an absolute minimum standard. The fallacy here lies in the supposition that there are no mechanisms to regulate population growth except death by starvation. In fact, population trends are closely geared to social and economic conditions. During the past century or two, Sweden has progressed from a situation of high birth rates and high death rates to a situation where at least medical progress and economic prosperity have helped to bring about a sharp decline in both trends. In many of the countries which are currently making the heaviest contributions to the growth of world population,

(1) For a review of the 'Limits to Growth' model of the Club of Rome, see for example:
- H S D Cole et al. (University of Sussex): "Models of Doom - a Critique of the 'Limits to Growth', Universe Books, New York, 1973, and 'Thinking about the Future', Chatto and Windurst for Sussex University Press.
- E Zwicker, "Moglichkeiten und Grenzen der modellgestutzten Prognose soziookonomischer Entwicklungen: Dargestellt am Beispiel der Weltmodells von Meadows und Forrester", Preprint 1/76-84, Int Inst of Management, Berlin
- Nathan Keylitz, "Understanding World Models", IIASA, April 1977
- Sam Cole, UN/UNITAR "Global Models and the New International Economic Order", draft 1977
- Christopher Freeman/Marie Jahoda (editors), "Word Futures - the great debate", Martin Robertson, London 1978.

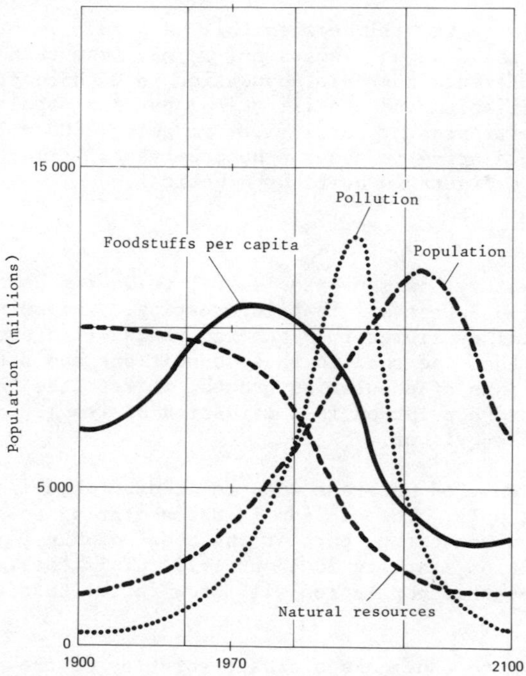

Figure 29. The future of the world - a model of doom

Population, food production and industrial output all grow exponentially until the rapidly dwindling resource base precipitates cuts in output of foodstuffs and industrial goods, (cf. the declining curve for natural resources). Owing to natural lags in the system, both population size and pollution effect will go on increasing some time after industrial output has attained its maximum. Population growth will eventually be curtailed by a rise in mortality rates due to a decline in the per capita availability of foodstuffs and medical services.

Source: D H Meadows, D L Meadows, J Randers & W W Behrens III: The Limits to Growth, Universe Books, New York 1972. The above diagram is taken from H S D Cole et al. (the Sussex group), "Models of Doom", New York 1973, and "Thinking about the Future", London 1973, which includes a discussion of this world model.

mortality rates have declined while birth rates are still high. But there is no reason to discount the possibility of these countries also reducing their birth rates as Sweden has done already.

In fact gross reproduction is already expected to decline during the 1970s in all regions except Central Africa, where the decline may come later.(1)

No informed observer of population growth believes that it can go on indefinitely, or even for another century. The moot points are when and why population will

(1) United Nations World Population Conference, Bucharest, 19th-30th August 1974. Report, "Recent Population Trends and Future Prospects" (E/CONF 60/3).

stop growing and how large it will be when this happens. In order to have a rough figure in mind during the ensuing discussion, we may note that one of the background studies for the UN Conference on World Population in Bucharest, illustrating a possible state of equilibrium and stabilization, put the population of the world at about three times its present level (i.e. roughly 12000 million people). If so, this situation would arise in about a hundred years' time. It must of course be emphasized that this figure is quite hypothetical.

Foodstuffs Per Capita

In fact question of exactly how many people will be living on the earth in fifty or a hundred years' time is not all that interesting. The main concern should be that the people who will be living in a hundred years' time should do so in fairly decent conditions and that the road to those conditions has not been lined with disasters. In discussions of population growth, it is often taken as understood that the very size of the population is a yardstick of its living standards. But this is by no means self-evident.

Size of population is only of interest when related to some indicator of living standards, e.g. food supply. The decline in the amount of foodstuffs available per capita does in fact constitute part of the model of doom pointing to a collapse in the supply of nutrition starting in about 1970. This thesis of collapse is based on the assumption that population will grow faster than the potential to produce food.

It is not possible here to undertake a closer scrutiny of the premises leading to the conclusions drawn by the Club of Rome. Suffice it to say that our opinion concerning food potential is very different from that underlying the Club's portrayal of a collapse.

A certain change in eating habits and a certain reduction of calory intake would not impair health standards in the industrialized countries. Furthermore, as we saw in the last chapter, there is a potential to expand global food production. The main question is whether it will be possible for this potential to be mobilized and on what social and economic terms this can be accomplished. This in turn leads onto the question of how the resources produced are to be distributed.

Natural Resources

The model of doom is founded on a static view of the availability of natural resources. No matter how great they are said to be, therefore, they inevitably run out at a given point in time.

As was pointed out in Chapter 2, it is wrong to think of resources as being given once and for all. True, the natural resources we are utilizing today are obviously limited. This applies equally to those which are renewable and those which are not renewable. The renewable resources are subject to physical limits of area and biological production capacity. But these assets can be increased by better utilization of already cultivated areas (better yields and adjustment to local conditions), and by utilization of assets not very commonly used at present (e.g. algae and deep sea squid etc. for food, fast growing stocks of sedge and suchlike for food or energy inputs).

The situation applying to what are called non-renewable resources varies a great deal from one substance to another. There are substances which are so abundantly available in the earth's crust or in the sea that their physical existence does not constitute a limiting factor. This applies, for example, to iron, aluminium

A highly industrialized meat factory in the United States. The farmworkers tending the livestock in these vast acres need vehicles to help them cover considerable mileage along the service roads between the fields every time they drive round with the cattle feed. (Picture: Manufacturers Hanover Trust Company).

The scene pictured below, with the shepherd leading his flock up to the mountain pastures, is obviously more pastoral. But in the developing countries too, extensive acreages are beginning to be used for raising livestock on an industrial scale. (Picture: Yngve Baum, SIDA).

and magnesium. Where substances of this kind are concerned, the limiting factor is more likely to be the amount of energy required for their extraction and purification. There are other substances whose availability is such that they may become too scarce within a century to be used except in extremely limited quantities.

It is difficult to make an exact estimate of the long-term availability of many of these rarer substances. New methods may broaden the base, both through discoveries, new extraction methods, (for deep-lying ores or those whose composition prevents them from being viable today), or new locations (ocean deeps, the Antarctic).

Possibilities of other kinds are also conceivable. If and when a process of industrialization combined with social and economic progress gets seriously underway in the developing countries, their output potential, which at present is underutilized, will probably do more than cover a steep rise in their own consumption of minerals. They would at the same time be able at least to maintain their level of exports. In a situation where the developing countries were more on a level with the industrialized countries, they would not have cause to fear - as they often rightly do today - the exploitation of their mineral wealth in the interests of other countries rather than their own.

The importance of technological development and substitution in the resource and raw material sector up to the present day is clearly illustrated by an American study to which we referred briefly in the previous chapter.(1) In that study, an investigation was made to see whether the scarcity of natural resources had restrained the economic growth of the United States between 1870 and 1960. It was theoretically conceivable that the cost of extracting a unit of a natural resource, e.g. one ton of metal, would show a regular increase as progressively inferior deposits had to be utilized. This hypothesis was not borne out by the empirical inquiry.

The authors concluded that production techniques had hitherto improved to such an extent that anticipated tendencies towards rising costs had been successfully eliminated (and even reversed). The economic system has possessed a great capacity for 'adjustment' in the course of a hundred years. Growth has been maintained even though natural resources have tended to become gradually less and less available.

Thus it is clearly a mistake when dealing with future perspective to regard the system of natural resources as completely rigid and beyond control or influence. In this respect, among others, the premises of the model of doom are unrealistic. On the other hand, historical experience is no guarantee that technological progress will continue at the same rate as hitherto.

Greater recycling and substitution of more abundant resources may also contribute to solving the problems of raw material supply. It is on account of possible adjustment mechanisms of this kind that the exhaustion thesis of the model of doom is unduly rigid. There is nothing indicating that society is fettered to the raw materials that are being used today. The American researchers H E Goeller and A M Weinberg (2) have outlined a two-hundred-year course of development in which society gradually switches over to a resource base comprising the few but more commonly occurring substances in the earth's crust. Their analysis of the materials demanded by social services of different kinds show that no substance which is absolutely indispensable is in danger of actually running out for the next few

(1) H J Barnett & C Morse, "Scarcity and Growth", Baltimore 1963.

(2) H E Goeller and A M Weinberg, "The Age of Substitutability - or what do we do when the mercury runs out?", publ. in M Goldsmith et al. (ed.), "A Strategy for Resources", North Holland Publishing Co., 1977.

centuries (given present-day extraction rates). But this long-term solution to the question of resources for humanity will demand technological progress to facilitate a transition to the more commonly occurring materials, and it will also demand a supply of energy that is not unduly expensive, because the new materials cannot be produced without large quantities of energy. Despite their fundamental optimism, the authors state quite frankly that the necessary process of adjustment can be extremely wide-ranging and difficult, even if a solution is found to the energy question.

Pollution

The Barnett and Morse study of the cost of extracting one unit of a natural resource indicates a constraint which is indirectly bound up with the extraction of resources, namely environmental effects. The authors believe that increasing environmental side-effects may eventually prevent technological progress from reducing unit costs as it has been able to do so far.

We also regard environmental effects as a very important constraint on man's future extraction of raw materials. We shall be returning to this point presently. We object to the picture conveyed by the model of doom, not because we consider the problem unimportant but because the really serious threats do not correspond to the descriptions given in the model. Firstly,the model assumes that society will ignore pollution until its effects on production and human longevity become disastrous. But society and individuals are hardly likely to be that passive.

Secondly, there are very many different types of pollution. Some of them can presumably be dealt with quite easily; this applies, for example, to certain types of toxic emission from factories and similar facilities. Countermeasures are less easily accomplished when the pollution is itself an active component of a form of activity and the latter is very widespread. Control at source may then be difficult if not impossible. The use of nitrogen fertilizers in agriculture is a case in point. Nitrogen is hard to dispense with, and nitrogen run-off from farmland to water courses and groundwater is hard to control because the use of these fertilizers is so widespread.

In the third place, the pollutants which become harmless as quickly as the model of doom indicates are perhaps not the gravest ones. The most dangerous pollutants are probably those which gradually accumulate. It can take a very long time for substances of this kind to be degraded or eliminated from the biosphere. We know very little indeed about the long-term effects of even small concentrations of different substances. The combined effects of different substances discharged by industry into the various natural cycles may also be rejudicial to biological organisms, man included.

It should be noted that energy production, like the availability of many substances also has an important bearing on the environment. Perhaps the gravest long-term threat posed to the environment by the production of raw materials may be the indirect effects of energy production. Here we need only refer to the possible climatic effects of discharges of carbon dioxide and to the problems entailed by the necessary dischargesof waste heat.(1)

Questions concerning what are termed the outer limits to human activity are destined to become very important in the long run. There is a limit to the demands which the biosphere can accommodate where human activities are concerned.

(1) Carbon dioxide invariably results from the combustion of coal, oil, wood and similar sources of energy. Concerning the effects of carbon dioxide, see e.g. Bert Bolin, "Energy and Climate" (the Energy and Society project group, the Secretariat for Future studies). (In English).

Pollution resulting from the discharge of toxic substances into the atmosphere and water may become a serious and, ultimately, constraining factor where the extraction of raw materials is concerned. The citizens of Tel Aviv complained about the stench from the Yarkon River. The environmental experts investigating this pollution had to wear protective masks. (Picture: UPI, Frankfurt am Main)

"When these limits are transgressed, human wellfare - even life itself - is threatened. No social or economic endeavour, including the satisfaction of basic human needs, can safely ignore the restraints imposed by outer limits". [1]

Everybody agrees that these limits should not be exceeded, but too little is known about them today and great efforts will be needed in order to improve our knowledge. We also know very little concerning which limits are susceptible to influence and which are not. "The complexity of outer limits is such that it will not be easy to identify them. For many years approximations and general indications of threats to the outer limits will have to suffice, and prudent action must be taken on the basis of the available knowledge, while acknowledging its deficiences and the need to improve it". [2]

The future portrayed by the model of doom is not inevitable. In discussing the constituents of this model, it has not been out intention to paint a rosy picture of a future with no problems as being a likelier prospect. Instead we have tried to show that natural resources and growth have to be studied in a wider context of social organization in order that the threats be evaluated. Furthermore, as we observed in the introduction to this chapter, we have only a limited knowledge of relationships and fundamental data. It is hard to identify the factor or the combination of factors by which future developments will be determined. Opinions concerning many of the objections raised here to the 'model of doom' vary a great

(1) *UNEP Governing Council, Fifth session 9th-25th May 1977. The Environment Program - Report of the Executive Director*, p. 27. (UNEP/9C/90).

(2) *Ibid*

deal from one person to another. This is why different observers can come up with such completely different total pictures of the future.

Most people would agree, however, that <u>the next hundred years will bring a number of distinct changes to the development processes which have operated so far</u>. This applies both to population trends and to the type of resource base which we will have to start using. Comprehensive changes of social and economic structure may occur for reasons which are not directly linked with natural resources. This in turn may influence the circumstances in which resource problems have to be tackled.

Most of the approaches to a solution which have been outlined as an alternative to the model of doom have emphasized technological progress, as reflected by production techniques, developments of substitutes and environmental engineering. Obviously some form of technological progress will be needed in order to assure the future inhabitants of the world of a supply of resources. The question is whether this progress is attainable in the first place, what direction it ought to take and what expense (in the broad sense) will be entailed by it. We shall now turn to consider these questions very briefly, one by one.

<u>Will technical solutions be found?</u>

A possible future shortage of natural resources can have various degrees of impact on the community as a whole. Its consequences may be fairly slight if it does not jeopardize vital social services or if it can be overcome by switching away from scarce resources to other resources. If the alloying elements used in the manufacture of stainless steel run out, technological progress may allow other methods of protecting iron against corrosion to totally replace them. Alternatively other substances (e.g. aluminium) may come to be increasingly applied to purposes of this kind.

But the shortage of resources may assume more serious guises than this. A certain resource which becomes scarce may be of extensive direct or indirect importance to many other resources. This will be particularly momentous if the resource which runs short is not substitutable. For example, if there should be a shortage of phosphorus - which does not seem likely at present - grain production would be disastrously affected. A shortage of water in relation to the demands of a greatly expanded agricultural sector might be taken as another example.

At the outer end of the scale of menaces, we have the situation where technology has not succeeded in converting the present-day resource base into one which is founded on more commonly occurring substances, i.e. efforts to prevent a more general shortage of resources have been unsuccessful. Moreover, if the worst comes to the worst and environmental engineering has failed, the biological foundations of production may have been destroyed on a global scale. These last-mentioned doom situations are probably purely theoretical. Before such a situation can arise, profound changes (e.g. war) will probably have supervened to transform the circumstances completely.

Shortage of energy is one of the gravest impediments to adjustment to a new resource base. This is not the place for a discussion of long-term energy supplies. The important point is that a solution of resource problems generally cannot be viewed in isolation from the need to obtain energy. Any technological optimism must therefore incorporate a fairly bright view of the prospects of solving the energy question.

Scepticism concerning technological advancement being able to cope with the future resource problems is partly rooted in a feeling that although the innovative technological potential migh be kept constant the emerging hazards and side effects from technological solutions are growing. Thus according to this view a limitation of the use of technology will be necessary in the future.

For some of the sceptics there is even doubt about technological innovation to continue in the future at the same rate as it has done in the past - a sort of 'limits to innovation'. According to this view there is real doubt whether there exists any possibility at all of achieving the future technology at a rate which can offset the decline in the metallic content of minerals once the higher grade deposits have been worked out. The existence of a historical sequence of this kind does not mean that similar developments can go on indefinitely. It is very likely, the argument goes, that somewhere along the line this development potential may be exhaused. If so, we ought not to plan as if uncertainty implied that solutions were assured. After all, who would set off to fly the Atlantic with only enough fuel for half the journey, comforted by assurances that a new type of fuel was sure to be discovered en route?

Whereas the sceptics take the element of uncertainty as the foundation for a reserved attitude towards technology, the advocates of technology argue that the future is bound to imply an element of uncertainty, especially regarding discoveries which have yet to be made. In their view, it is self-evident that something will be discovered, and this something, they maintain, will in all probability exceed all the prospects held out by our technological forecasts. In support of this argument they plead the record of historical experience, which tells us that long-range technical forecasts have very often turned out to be underestimates. Forecasters are afflicted by a lack, not a superabundance, of imagination. This is natural, because the accumulated imagination of many people over a long period of time is the very essence of research and development.

The pessimistic school, on the other hand, maintains that the uncertainty surrounding the technology of the future includes both the mertis and demerits of future opportunities. It can therefore be exceedingly dangerous to let the genie out of the bottle so long as the price to be paid remains unknown. Fear of the side effects concealed by our uncertainty prompts a restrictive attitude.

What Sort of Technology?

The support which the community gives or denies to different technical projects is influenced by the manner in which technology is interpreted. A situation may occur in which this route to general development and the creation of resources is rejected, at least in part. The manner in which investment in new technology is supported - if at all - by the population at large is therefore a factor of vital importance. Developments may move towards great specialization, towards processes of greater complexity, with the result that a limited number of experts are the only people in command of technology. A society of this kind would exhibit a variety of dangers, including the risk of technocracy and of powerlessness on the part of the many. (In Sweden, there is already such a debate concerning e.g.computer technology). Alternative methods of technological assessment are being discussed today. Technology evolved in connection with a broad public debate and with direct consideration for social and ecological effects could result in new techniques being more readily accepted and more efficiently utilized. Therefore the question as to whether technical solutions will actually materialize and be used cannot be divorced from the ability of the individual to improve his knowledge and find an outlet for his initiative.

In this way technological progress will be influenced - if not determined - by the general course of political, social and economic development.

There is increasing talk today of 'appropriate technology', 'small scale technology'

Modern technology present many opportunities of broadening the resource base and keeping raw material costs down. Satelliteborne remote sensing cameras will in future provide a more efficient means of, for example, monitoring global harvests and diseases affecting timber stocks, and it will also be possible to use them in the search for new ore deposits. But technology can become a menace unless the knowledge possessed by the majority of the population grows in step with the introduction of new techniques. This is important not least where the developing countries are concerned. Forms of technology control are therefore no less important for the future than the actual development of the new technical components. This picture was taken at the offical opening of a satellite telecommunications unit in the United Arab Emirates in 1977. (Picture: Deutsche Presse Agentur GmbH, Frankfurt am Main)

---

and so forth. These expressions betoken an increasing realization that different situations demand technological solutions of different kinds, even though the problem in purely technical terms may be one and the same. This is particularly the case regarding the technology which the developing countries may feel that they need. It is not self-evident that they should inherit solutions evolved by the industrialized countries in a different social situation.

Costs

If we assume for the sake of argument that problems of resources occurring during the next fifty or a hundred years can still be solved by means of technological progress, this does not imply that no sacrifices will have to be made by society or its individual members. In many cases there costs will be obvious. For example, they may take the form of stricter rules concerning waste management, or they may take the form of even more large-scale production methods. In some cases the changes involved may appear as costs to some people while others may actually view them as positive steps towards a better quality of life. Changes of this kind could include the abolition of private car ownership, increased circumscription of the landlord's right to the enjoyment of his property, and so on. In these cases, priorities could be made the subject of a political discussion in which benefits and costs can be presented and balanced against one another.

Risks as an Element of Cost

Things will become more difficult to assess if gains in the form of additional resources have to be balance against risks associated with technology. All we can do here is to point to some of the types of risk already referred to in connection with questions of resources.

All the countries of the world have become increasingly interdependent, economically and technologically. It is arguable that their ecological interdependence has also grown. Certain forms of dependence can have substantial advantages. The local failure of a harvest nowadays need not spell starvation as it so often did in the past. At the same time, 'globality' implies a hazard due to (for example) the ability of ecologically destabilizing factors to spread very rapidly throughout the world. We have already referred to one example of this kind, namely the hazards associated with an increasingly narrow genetic base for foodstuffs; a plant disease can suddenly spread all over Eurasia. Thor Heyerdahl [1] maintains that the whole of the South Atlantic is already dotted with lumps of oil. If there were a succession of blow-outs, and indeed even with the present rate of spillage from oil rigs and shipping, this large scale pollution could have serious effects on fish and algae.

Many of the conceivable future efforts in the resource sector can be expected to augment the resource base at the price of greater risk. For example, the extraction of minerals from the seabed can involve risks to the marine environment. The use of technical aids to influence climate for the benefit of agriculture in one particular area might be detrimental for another area. There is also the use of new types of crops or timber which promise higher yields but have not been previously tested in their new surroundings. Many countries may opt for a short-term advantage even at the long-term risk of denuding their landscape of topsoil; for example, large-scale clearance of tropical rain forests is a case in point. A decision of this kind may emanate from ignorance, private or collective hunger for profits, or a desperate struggle for survival. In many such cases the effect has been to increase the probability of reverses, as well as to increase the gravity of their consequences should they occur.

Many of the resource technologies of the future will force policy-making bodies, one way or another, to adopt standpoints on questions concerning the 'outer limits' of the biosphere. In some cases, the outer limits and their exigencies may conceivably impose restraints on the production of raw materials, but it is hard to tell at present where this might occur. The balance between greater risk and increased supplies or resources will constitute an important strategic choice in the future. This balance will probably have to be struck to an increasing extent through the medium of supranational bodies, due to the risk affecting several states or the entire human race.

System Sensitivity as a Cost

The risk outlined above was more of a technical and scientific nature. The social

(1) known e.g. for his Kontiki and Ra sailing trips across the oceans.

systems of which raw materials form a part are also sensitive. Vulnerability to intentional or inadvertent interference can be expected to increase if developments continue on their present path.

Today we can already see how computer technology links different activities together. The breakdown of a single computer can paralyse large areas of Sweden. If energy cannot be produced in sufficient quantities, other raw material industries will be inhibited. The network of trade links is influenced by political judgments which are not directly concerned with trade. The social, economic and political systems are interconnected, and the sensitivity of society to a disruption occurring in one single sub-system increases accordingly. At the same time there is a danger of the vulnerability of the community being taken as a pretext for not tackling social problems.

Costs in the Form of the Conceivable Impact of Technology on Social and Economic Structure

It seems clear that increased extraction of resources brings greater uncertainty concerning the potential social effects of different technological solutions. Uncertainty of this kind is in itself a cost item. What, for example, will be the effects of different solutions on such variables as income distribution,co-determination in industry, or stress from the point of view of the individual?

Knowledge is might. If large-scale high-technology projects grow even more important, this can lead to an increased division of the members of society into those who are up to the mark and those who are not. Specialization and the need for expertise have reserved the discussion on many questions which are intrinsically political to small groups of decision makers.

Control of raw materials technology in a world where such technology is gaining in importance could put the groups of nations already possessing such technology into firm position of superiority. Increasing pressure on scanty resources might also make those in possession of technology less predisposed to share their knowledge with others. This could potentially be a factor running counter to the long-term equalization of living conditions in the world.

The interconnection of technology and social organization is an old question. There need not necessarily be a self-evident connection between technological and social organization, but there is a great deal to suggest that certain types of technology involving great hazards or operations on a considerable scale, for example, can obstruct the development of a democratic structure of decision-making.

Costs in the Form of Reduced Quality of Life

Many raw material sectors involve environmental costs which people notice and feel immediately. Greater efficiency in the forestry sector has led to a large-scale clearcutting technique which has come to be increasingly challenged, not least on grounds of ecological impoverishment. Possible effects of this kind, like the general defacement of the landscape, can be said to constitute an impairment of the quality of life. Extensive mining operations in populated areas can lead to the relocation of an entire community.

A situation may conceivably arise in the future where extensive populations are moved according to a plan, so as to make it possible for specific locations of resources to be used. It is reasonable to suppose that activity of this kind will be made conditional on the restoration of the landscape in one form or another after the minerals have been extracted (a practice already current, for example, in Germany, when coal seams are worked out). Even if the landscape is successfully restored, migrations and loss of the right to live on ancestral land can be regarded as impairments of the quality of life; cf. the question of migrations necessitated by the damming of lakes for hydroelectric power production.

Environmental changes, particularly those involving the destruction of ancient monuments or changes of vegetation, are a similar type of loss, involving the quality of life. Impact on a distant area which has been earmarked as a nature reserve is another but similar type of consequence which can be termed a loss to the quality of life, since it is liable to endanger future recreation opportunities or the possibilities of scientific research.

The utilization of new kinds of resources can involve greater occupational hazards (work on the sea bed is one such example), or it may demand work in remote places (such as the Antarctic), where conditions are generally inhospitable. Inability to settle in one and the same place for any considerable period is another type of cost which may increase.

We could go on to outline various other examples of losses of quality of life. In the long run, as the conditions involved in obtaining resources become more and more exacting, there seems to be a general likelihood of goal conflicts of these kinds becoming intensified.

To many people the mere physical fact of exploiting nature is in itself a loss of quality of life which has to be balanced very carefully against the material advantages gained. The following passage from a poem by the Swedish poet Goran Sonnevi [(1)] expresses the despair that is caused by technology blindly engulfing important qualities of life.

> On the contrary, it seems
> as though we
> by pollution, brutal
> consumption
> have imposed new and
> unprecedented limits
> on the expansion of
> the unknown
> We now have the power to stop
> the evolution of all species!
> We do not know yet how and where these
> limits will be hit
> We know they hit

It is not unreasonable that discussions concerning the extraction of resources in the future will have to focus more on such a feeling that something elusive, but nonetheless important, is lost in the process.

Greater attention needs to be paid to the different forms of cost entailed by the exploitation of raw materials. This applies equally to those forms which are easily defined and those which are more complex or obscure. It is not impossible that costs of this broad kind may turn out to be the real limiting factors, together with certain absolute restraints posed by the outer limits, rather than a direct physical shortage of individual substances.

As a rule, the aspects which we have here referred to broadly as 'costs' are not included in present-day calculations of the net benefits of new technology, previously exemplified by the Barnett and Morse study. The aspects which are not

(1) Goran Sonnevi, "Det omojliga", Bonniers, 1975

measured in the market place and which are concerned with the overall maintenance of life can, as we have seen, be 'consumed' (or at least impaired), just like 'ordinary' natural resources. This is particularly important when it results from the use of new, 'resource-expanding' technologies.[1]

The demands of a future resource situation may eventually lead to a transformation of our present-day economic and political organization. It is obviously impossible to foresee what the new forms will look like. On the one hand, the pressure exerted by an austere resource situation may lead to authoritarian forms which we would not accept today. But on the other hand, the resource situation may demand a mobilization of the creative powers of all members of society, in which case a social pattern may be required which will offer creative potential more scope than it enjoys at present. This would be a powerful argument in favour of a deepening of democracy. Seen in such a perspective, the system of decision-making would have a vital bearing on the general pattern of future development. It is of their very nature that these great issues cannot be resolved for all times. It is the responsibility of each new generation - and in the present chapter we have been looking two or three generations ahead - to ensure that the democracy which has been achieved is not lost, but can be further developed.

(1) J V Krutilla, V K Smith and R Kopp, "Natural resource availability, environmental quality and economic well-being", Resources for the Future, 1977.

6 Sweden and Raw Materials

Sweden's raw material profile

Sweden is a highly industrialized country. Her production is partly based on raw materials of her own, above all foodstuffs, raw timber and certain ores, and it is also within these sectors that we find the traditional export commodities, the natural basis on which Sweden's industrial expansion has been founded. Post-war developments have increasingly integrated Sweden into an international system of trade in which raw materials, semi-manufactured products and finished products move from one country to another at different stages of processing.

One way of illustrating the functions of raw materials in society is by means of flow charts depicting total raw material utilization. The flow incorporates a succession of stages from primary industries (farming, forestry, fisheries, extractive industry), via processing industry (e.g. pig iron and steel mills) to final manufacturing (e.g. the engineering industry) and, ultimately, waste management and recycling, if any.

Material flows can be used for various purposes. They can constitute an important input to economic analysis, by revealing how price increases for raw materials are disseminated throughout the economy. Value flows are then used. From the standpoint of environment policy, or with reference to transport requirements, material flows by volume are often just as interesting as material flows expressed in values.

Figures 30 and 31 represent two overall material flows for Swedish imports and exports in 1974, in tons and US dollars respectively. It shows firstly that Sweden is a net exporter of raw materials in terms of volume. Iron ore, timber, pulp and paper, measured in tons, account for 90 per cent of Sweden's exports. The biggest import items are mineral oil and petroleum products, which add up to about two-thirds of the volume of imports.

The second fact which can be deduced from figure 30 is that the outgoing flows represented by exports and waste do not correspond to the in-flow. This reflects the accumulation by our society of an increasingly large apparatus of production, distribution, transport and durable consumption goods. The greater part of physical accumulation comprises roads and housing, i.e. sand and gravel.

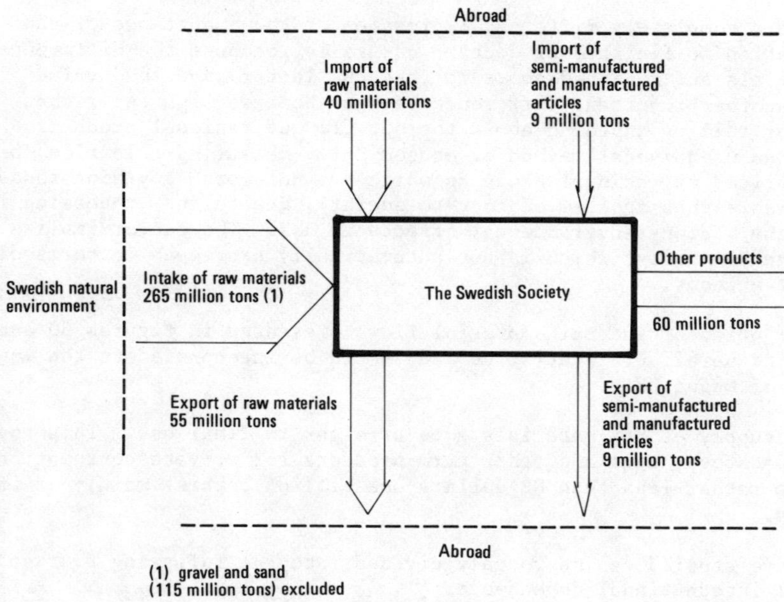

Figure 30. The flow of materials for Sweden - in metric tons

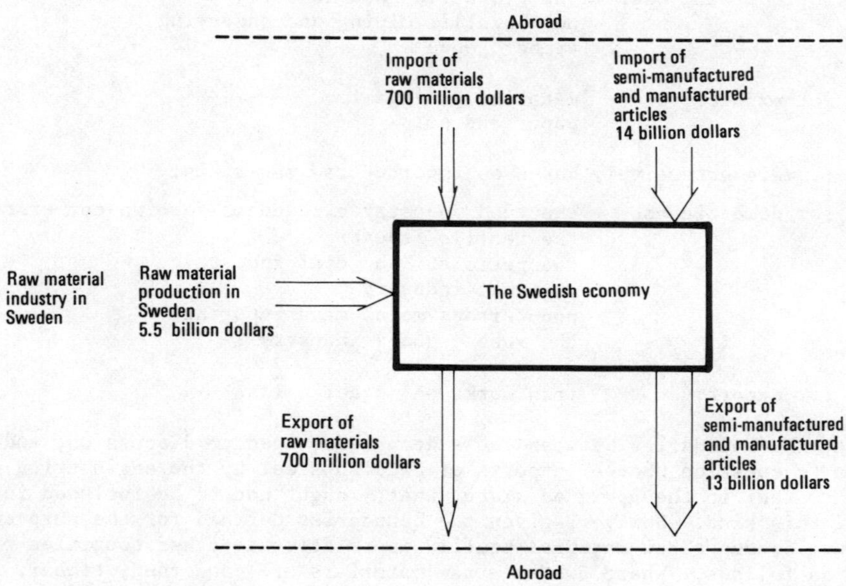

Figure 31. The flow of materials for Sweden - in US dollars

Figure 31 conveys a completely different impression of the importance of raw materials in relation to finished products - naturally, because the difference between raw materials and finished goods is that the latter have had value added to them. Another important difference between the two figures is that figure 31 does not tell us anything about the out-flow of residual products. There is in fact no unequivocal method of economically measuring this flow, because waste has been defined as residual products which are not worth anything today. Obviously waste represents an item of cost to society, due to the processing it requires and to the various environmental effects it has. The latter implies expenditure by the community, which is not automatically expressed in terms of value in a market economy.

The general description of the main material flows presented in figures 30 and 31 is founded on quite unreliable statistics and should be interpreted at the most as illustrations of magnitudes.

One-third of the supply of raw materials goes straight to final use. This applies, for example, to potatoes, eggs and other farm products for private consumption. Exports totalling rather less than US dollars one billion consist mainly of iron ore and roundwood.

The sectors we have studied can be roughly divided into the following four groups in terms of total international dependence.

A. Sectors based principally on domestic raw materials:

1. for domestic use: the protected food industry
   non-metallic mining and quarrying
   timber goods

2. for export: metallic mining
   paper and pulp

B. Sectors more extensively based on imported raw materials:

1. for domestic use: the food industry exposed to foreign competition
   the textile industry
   the petroleum and coal industry
   chemical industry
   non-ferrous metal manufacturing
   the rubber goods industry

2. for export: iron works and steel mills

Of course, the boundaries between these groups are open to discussion, and it is also open to question whether exports of Swedish steel by the engineering industry - steel in the upgraded state, that - ought not to be included in a review of this kind. However, given the boundaries defined for the purposes of our review, Sweden's raw material profile is relation to other countries can be outlined as follows. Where Sweden's raw materials are concerned, timber, iron, zinc and lead are still important export commodities. Exports of grain are perhaps also to be included in this category. Most of the non-energy raw materials which we import are processed for use within the country. In the case of steel, Sweden constitutes a processing station for imported alloying elements and imported coal.

Sweden and raw materials - the scenario work of the RoR study

The purpose of the RoR study has been to investigate Sweden's long-term supply of resources and raw materials. A great deal of the work done in the course of the study has also been devoted to efforts to find out what can be said about Sweden's future raw material requirements and about possible limits to the supply of raw materials.

There have been three points of departure for this work. Firstly, the chronological perspective has been confined to the years up to 2000. This was because it was considered that a perspective of 10-25 years can be perceived without very large changes in the overall structure, e.g. patterns of consumption and the composition of foreign trade.

The second point of departure of the forward-looking activities of the RoR study was that Sweden's long-term supply of raw materials was greatly dependent on changes in the world at large, changes which Sweden cannot influence to more than a very limited extent, politically or in other respects. The various possible lines of development which have been deduced on the subject of Sweden and raw materials have therefore been based on variants of hypothetical "international environments".

The third point of departure has been an attempt to indicate the similarities and differences between various possible but not necessarily probable lines of development by working out different perspectives. It has not been our intention to arrive at a forecast, i.e. to try to say what is most likely to happen.

The RoR study has employed two methods of outlining alternative courses of development. One of these has been based on the econometric model employed by the latest Swedish economic long-term survey for the prospect up to the turn of the century. The alternative descriptions of the Swedish economy generated by different hypotheses regarding the international environment were fed into the model used by the National Swedish Board of Economic Defence (OEF) when planning emergency stocks. The results of these studies gave the consumption of certain strategic raw materials in Sweden.

The second method has involved the writing of scenarios for Sweden's supply of raw materials. A scenario is a description of a possible future situation usually accompanied by a possible course of development from the present state of things to the situation described. Outwardly speaking, a scenario may be very short and deceptively simple, but it is in fact based on a number of assumptions concerning the workings of present-day society and the identity of the most important factors underlying social change. A scenario need not necessarily focus on a situation regarded today as highly probable.

The distinctive feature of scenarios is their form, which is directly descriptive and constitutes a deliberate albeit simplified attempt at studying the whole without losing sight of the interfaces between different sectors of society or different academic disciplines. The simplification usually has the effect of making the statements in the scenario look like generalizations (because a short passage of writing cannot accommodate many reservations). The overall picture is given in a rather impressionistic manner. But the cardinal virtue of this direct, concentrated presentation is that it focuses on certain essential features of an otherwise complex description. In other words, some of the trees are probably forgotten, but one sees the whole of the forest clearly.

Our efforts at composing scenarios can be summed up as follows. The first round of draft scenarios comprised broad, generalized political descriptions of "what happened" in the world, i.e. events occurring in individual great powers and other countries during the next twenty-five years. Perhaps the result made entertaining reading as a projection of certain present-day problems, but above all there was one obvious deficiency; it was very difficult to establish a tenable link between the international surroundings thus depicted and our real subject, which was Sweden's long-term supplies of raw materials.

The next round of scenario work therefore constituted a reactions to the experience of the first. This time we started with Sweden's supplies of raw materials and with individual raw materials of interest (grain, chromium, manganese, tungsten, etc), and we studied the problems which might come to affect Sweden's supplies.

One obvious problem entailed by these scenarios was that they presupposed rather different international contexts. In a scenario, it is easy enough to create a supply crisis for certain raw materials if you are a full liberty to make whatever assumptions you please concerning international developments.

The third round of scenario work involved trying to compose a more complete scenario aimed at covering the most important aspects of the world at large, as well as general developments in Sweden. A more ample presentation of this kind would make it possible to do the following:

(a) to elucidate assumptions which would otherwise be hidden or not expressed;

(b) to reduce the possibility of arbitrary choices of the authors within the given framework of raw material scenarios;

(c) to establish consistency between different raw material scenarios.

These efforts resulted in a detailed but rather loose-knit picture of prominent tendencies at the global level and in various fields - politics, economics, technology and ecology - together with their possible impacts on the raw materials sector. It contained clear points of intersection which showed that all the tendencies we had included could not continue "as usual"; structural changes of some kind were bound to occur.

The outcome, however, was not a success. The aim of creating a framework for discussions of the raw material sector was not achieved, because the frame did not hold together, i.e. the tendencies in the various areas or sectors were incompatible and therefore often failed to produce consequences applicable to the discussion concerning raw materials.

The fourth and final round of scenario work was aimed at a synthesis of this experience. The approach we arrived at consisted of four short world scenarios, each of which describes very different international contexts. These were characterized by the following: supranational co-operation aimed at promoting the economic and social development of the Third World countries; increased but selective co-operation between groups of industrialized countries and Third World regions, e.g. Europe (the EEC) and Africa; a continuation of the present international division of labour; and a world dominated by a conflict between industrialized and developing countries.

In the light of these world scenarios, it would have been possible to compose scenarios for individual raw materials (iron ore, grain, timber products, etc.). The questions asked in each scenario (not least as regards effects on Sweden) could be expected to vary depending on which world scenario was allowed to call the tune, as it were. It was not intended to discuss all raw materials in the light of all world scenarios. One of the points of the whole approach was that it would enable us to select questions and situations of particular interest from a Swedish viewpoint.

The scenario work of the RoR study was shelved before the raw materials scenarios had been fully evolved. This was partly because in the course of our work we had grown increasingly sceptical of the method, or at least of our ability to use it in such a way as to obtain results that would arouse interest and provide food for thought. The very form of the scenario made quite heavy demands on the authors at the same time as the usefulness of this method - e.g. discoveries of hitherto unobserved problems, points of intersection between different tendencies - was quite limited.

Two perspectives on Sweden in around 1990

Another line of investigation was provided in the form of an econometric model. The econometric model which had been devised for the 1975 economic long-term survey of Sweden was used to illustrate the effects of different world environments on developments in Sweden.(1)

Assumptions were formulated concerning two alternative world environments (schematically termed conflict and co-operation respectively) which could be used in the model. A series of computer runs yielded highly detailed particulars concerning GNP development, consumption and investment, employment and output in different sectors, and also concerning foreign trade. With the aid of the model used by the National Board of Economic Defence for planning emergency stocks, it was also possible to arrive at a demand for a number of raw materials in the various perspectives.

"The Reference Picture"

The starting points of this scenario were the premises adopted concerning the economic situation in 1980 by the 1975 Government long-term economic survey. This forecast includes data about the labour force and productivity as well as the growth of the housing sector and the public sector. A reference picture of Sweden in 1990 was obtained by an assumption concerning the size of gross investments - it is assumed that they will grow almost three times as fast during the 1980s (5.6% per annum) as during the 1970s (1.9%) - and with an unchanged structure of foreign trade. This reference picture is presented in more detail below, parallel to the other persepctives.

Two remarks should be made concerning the reference picture. The point of departure is the situation assumed for 1980. This involves the assumption(2) that a balance of current payments has been achieved in relation to other countries. In view of the uncertainty surrounding the present state of the international business cycle, this balance may very well not be achieved until some time during the 1980s.

Secondly, the date 1990 should not be taken literally. It merely represents a future point in time which is sufficiently remote for essential structuralrelations to have loosened and for changes to have occurred in the balance between capital and labour in the production process. The situation outlined may therefore occur either later or earlier than 1990. Accordingly, developments during the 1980s may also turn out differently. The reference picture is actually not at all intended as a forecast but as a basis from which to compare different paths of development.

(1) This work was initiated and performed by dr econ Tomas Restad, the Ministry of Economic Affairs.
(2) According to "alternative II" of the 1975 Swedish long-term economic survey.

Alternative World Environments

The choice of alternative world environments has been based on the extensive international debate on questions concerning raw materials. Simplifying matters somewhat, the debate on raw materials in the United Nations, for example, may be said to be bounded by two fairly extreme assumptions concerning what desirable and appropriate.

A great deal of opinion formation - in Sweden as well as in other countries - has been aimed at establishing a more equitable world order through the supranational planning of production, the control of raw material markets and, as a result, the elimination of global inequities of income. One manifestation of this school of thought is to be seen in the RIO report (RIO being short for Reshaping the International Order), compiled by an international group of experts led by the Dutch economist Tinbergen and published in 1976. Among other things, the RIO report emphasizes a number of measures in the raw materials sector which would increase and stabilize the export revenues of the developing countries. Its basic assumption is that the nations of the world can agree on a programme of action ultimately leading to greater fairness and equity.

A completely different school of thought is founded on the assumption that the affluent countries will not voluntarily relinquish their privileged position in the world. According to this school, it is only by an active struggle that the developing countries can become economically independent of the affluent countries and their transnational enterprises. International trade, including exports of raw materials, is regarded as an important factor in the maintenance or even the reinforcement of economic dependence. This makes a reduction of trade and increased processing of locally-produced raw materials an integral part of efforts by the peoples of the developing countries to raise their living standard. Ultimately this would also mean that an increasing proportion of the raw materials of the developing countries came to be consumed on the spot and not in the affluent countries.

As a factor distinguishing between these two future perspectives, we opted for the ways in which the two strategies describe desirable patterns of international trade and of the international division of power. These alternatives were termed "conflict" and "co-operation" respectively. The two world contexts were then defined more closely and 'translated' into economic variables of the kind included in the model. First, the structure of international trade was adjusted to suit the reference picture. In the conflict alternative, the prices of raw materials imported from the developing countries have been presumed to increase. In the co-operation alternative, international development assistance has been presumed to rise to 2 per cent of GNP.

An initial study of the conflict alternative (below entitled conflict 1) indicated that private consumption in Sweden could only increase by 0.8 per cent annually if investments in Sweden were to rise as rapidly as in the reference picture. It is very doubtful whether such a low increase in consumption will be politically feasible if GNP continues to grow by 3 per cent annually. The scope for wage increases in conflict alternative 1 is only a little less than in the reference picture, i.e. appreciably faster than the estimated scope for private consumption (0.8% per annum). In other words, heavy increases in taxation will be needed to restrain private consumption. This alternative also assumes that exports need to rise even faster than in the reference picture, at the same time as the opportunities of exporting to the developing countries and elsewhere will be limited (assuming that our basic assumptions hold good).

For these reasons, a further conflict alternative (conflict II) has been devised. If the growth of investments is reduced to 0.8 per cent annually, this will make it possible for private consumption to grow by 1.6 per cent annually. At the same time GNP growth will decline to 2.3 per cent annually, and this will mean less imports, which in turn will reduce the need for exports.

Viewed in a ten-year perspective, conflict II seems more attractive than conflict 1. The external trade balance is easier to meet, and at the same time there is greater scope for consumption. In a more long-term perspective, however, conflict II will mean either a long-run acceptance of less economic growth and less scope for consumption, or else merely postponing the issue. The additional cost of imports would then have to be paid for by a reduction of consumption later on.

Four perspectives on Swedish development are thus presented in the following pages. One of them is a reference picture based on the assumption that no radical changes will occur in the conditions governing the development of the Swedish economy. Two of them are based on the assumption that the world environment will be characterized by confrontation between industrialized and developing countries, resulting in mounting protectionism. These two perspectives differ in their assumptions concerning the growth of investments and the scope available for consumption. Finally, the fourth perspective assumes a development of the world economy in the direction of supranational co-operation and a wider international division of labour.

The Results

If we begin by considering the development of the balance of economic resources, i.e. GNP and its components, there is a strikingly small difference between the various perspectives at first sight. This is partly because the growth of certain items has been determined a priority, i.e. outside the model. This applies, for example, to gross investments, changes in investments and public consumption. The differences between GNP growth rates is barely 1 per cent per annum or almost 12 per cent after ten years. This is a consequence partly of the method and partly of the assumptions made. It also implies that all four perspectives are located in the "middle" of what can be considered possible or likely, despite fairly "extreme" assumptions concerning the possible nature of the international environment.

However, the assumptions made have very distinct implications for the structure of the Swedish economy. The simplest way of showing this is in terms of employment changes (cf. Table 14). It should be noted that the assumptions common to all four perspectives include a decline in total employment, measured in working hours, of 0.2 per cent annually throughout the period under consideration The differences between various sectors are therefore of greater interest than the absolute changes arrived at.

With just one or two exceptions, employment in the basic sectors and industry is assumed to decline while the number of working hours in the service sector is assumed to rise. This represents a continuation of previous developments. But it is more interesting to see the impact on the various sectors of the assumptions made concerning the world environment. Agriculture declines most rapidly in conflict 1, due to the slow growth of private consumption. In conflict II agriculture suffers the smallest diminution of employment, due to the slower rise in investments. Thus, a substantially slower mechanization process in the agricultural sector leads to a form of production which is more labour intensive.[1]

(1) More detailed results are presented in the Swedish version of this report.

Table 14. Employment in 1990 Index: 1980=100.

	Reference picture	Conflict I	Conflict II	Co-operation
Agriculture	52.7	44.9	78.4	49.4
Forestry	74.5	79.2	97.0	70.8
Mining	65.1	90.4	102.0	49.4
Raw material processing industry	79.2	83.4	87.7	76.8
Engineering and building	100.0	104.1	81.7	100.0
Other industry	59.2	65.8	78.4	56.8
Services	106.2	104.1	108.3	108.3
The economy as a whole	98.0	98.0	98.0	98.0

Some long-term factors with a bearing in demand for raw materials

Substitution and Technological Progress

At their present stage of development, econometric models tend to disregard the emergence of new technologies and the possibilities of new manufacturing processes being substituted for old ones in the production system. They do, however, deal relatively exhaustively with the possibilities of substitution in the consumer sphere.

This is not a very serious shortcoming if the model is applied to short periods, e.g. one to five years. But where longer periods and individual raw materials are concerned the method is liable to result in mistakes regarding the growth of consumption.

There are a number of changes which could occur in the raw material base within the not too distant future. Here are a few examples.

The adoption of new types of feed in the farming sector, e.g. rape; artificial fertilizers and feed based on straw and pulp.

In the minerals sector, a gradual reduction of such traditional metals as copper, tin or silver in favour of other materials like aluminium, composite materials and ceramics (frequently materials based on what are termed industrial minerals).

These examples represent cases when a new raw material takes the place of an old one. But it is possible that completely new processes or functions will replace traditional ones and that this will affect raw material requirements.

If pulp produced by thermal/mechanical methods supersedes that produced chemically it will be possible to use a large proportion of the wood fibre (though at the cost of increased electricity consumption).

New metallurgical processes can substantially reduce the wastage involved in metal working, e.g. swarf.

National Policy

It is not hard to find examples of the impact of national policy on the raw materials sector. In Sweden, there are topical instances to be seen in the textile and clothing industry, the steel industry and shipbuilding. Many sectors, however, are just as dependent, if not more so, on the policy adopted in the outside world. This is an important source of uncertainty when gauging the future situation regarding raw materials.(This issue is further discussed in the next chapter).

Longer Life Cycle

Another type of uncertainty regarding future developments is connected with the way in which the tastes and consumption patterns of individual people may evolve. This applies, for example, to the strength of future demand for public services or the communal value of a clean countryside. In the ultimate analysis, this could mean the termination of consumption of certain raw materials in Sweden, irrespective of their price. All we can do here is to indicate a number of tendencies towards change in value structures which may substantially affect demand for raw materials over a twenty-five-year period.

Prolongation of the service life of consumer durables, for instance, reduces the need for virgin raw materials. If a car or a refrigerator is scrapped after twenty years' service instead of ten, this will reduce the demand on the raw material base. It would not necessarily halve consumption of virgin raw materials, however, because net consumption is also affected by the degree of recycling and by technological progress. The latter can often reduce the specific raw material requirement for a given product. (Present-day refrigerators, for example, weigh a good deal less than those which were being manufactured twenty years ago).

It is quite possible, however, that an increase in the service life of products will become a progressively more important argument in the manufacture and sale of products during the next twenty-five years, because ideas concerning product design and the present discussion of patterns of consumption, with its moralizing undertones, are unmistakeably departing from the throw-away ethos of the 1950s and 1960s. Apart from the influence of the general climate of opinion on the pattern of consumption there are two more arguments in favour of this type of development. Whatever the course of developments in the energy sector, it seems inevitable today that economising on energy will be given a prominent position among the social priorities of the future. Both for objective reasons - longer service life means energy saving in the great majority of cases - and for reasons of public opinion, this tendency will probably influence the structure of demand for consumer durables.

Secondly, the tendency towards greater partnership in industry (the trade unions play a more active role in long-term policy discussions of enterprises nowadays in almost all western countries) may very well result in priority being given to short production runs of high quality products (with the skill and craftsmanship they imply) rather than long production runs, uncertain quality and a fast working pace. This latter argument may, however, be regarded primarily as a hypothesis.

This does not necessarily imply that demand for raw materials in a medium long perspective will be substantially lower than it would have been without the presumed change in preferences. For example, the impact of longer service life on the consumption of raw materials may not become apparent until the end of this century owing to lags in the social and industrial adjustment process. On the other hand, the structure of demand can be expected to change in favour of raw materials which measure up to exacting qualitative requirements. It is also conceivable that the raw materials used for a given product will be adjusted more closely to the performance required of that product, so as to limit the unnecessary utilization of high quality resources.

Saturation Limits

Changing preferances can be reflected by other criteria than longer service life. One of the fundamental questions surrounding future industrial development concerns the extent to which material consumption can go on increasing in the most affluent countries. There are some areas in which Sweden quite definitely seems to have achieved saturation levels which are unlikely to be exceeded in the future. Calory intake per Swede is a typical example, and the Commission on Agriculture has forecast that it will stagnate or decline during the period ending in 1985 (cf table 15).

Table 15. Nutritional value of diet, kcal and g per capita per day, 1960-1985.

	1960	1972	1973	1985 (forecast) Alt 1	1985 (forecast) Alt 2
Energy, kcal	3023	2938	2955	2861	2955
Protein, g	72	73	73	76	73
of which animal protein, g	48	52	56	57	56
Fat, g	125	119	119	111	119

Source: Production targets and food reservies, Ds Jo 1975:2. (In Swedish)

Food consumption in Sweden, measured in calories or grammes of protein, cannot be expected to increase much more than the population. Are similar tendencies discernible as regards other products? It is hard to find any general saturation in consumption of consumer durables, but a saturation limit can be observed where individual commodities are concerned; see table 16. Nearly all households today have a vacuum cleaner, and so there is hardly likely to be any heavy growth in this sector.

Table 16. Percentage of households owning various kinds of electrical equipment.

	1964	1975
Electric cooker, hot plate	70	88
Refrigerator	75	95
Freezer	22	65
Washing machine (own)	29	52
Dishwasher	1	16
Vacuum cleaner	84	89
TV, black and white	76	57
TV, colour	-	43

Source: Bo Diczfalusy, Energy and income distribution, Energy and society, the Secretariat for Future Studies, 1976. (In Swedish).

Since so many people have already acquired products of this kind, the next twenty-five years hardly seem likely to bring such a rapid growth of material standards as we have experienced during the post-war era. The forecast of vehicle density compiled by the National Road Administration also points to the existence of some kind of saturation limit.

On the other hand, we should not underrate the chances of developments continuing at almost undiminished speed for the next few decades. In the first place, possession of consumer durables is closely bound up with income, as witness table 17.

There is a great deal of scope for increased consumption before we reach the levels which are considered desirable today by people in the top income brackets. Consequently, consumption of raw materials is a reflection of the distribution of incomes in Sweden, at least indirectly.

Secondly, increased leisure and rising incomes can lead to a rapid growth in the number of secondary homes and, consequently, a rising demand for a second set of consumer durables. Holiday comforts (refrigerators, dishwashers, radio and television) have been one of the themes of advertising in recent years.

Table 17. Ownership of certain capital goods, 1973-1975, by income groups- Sweden

Possession of:	Percentage of households in various income groups, thousands US dollars						
	0 - 2	2 - 4	4 - 6	6 - 8	8 - 12	12 - 20	20 -
Dishwasher	1,8	2,4	4,7	8,3	16,9	31,8	60,0
Colour TV	9,3	24,1	32,8	41,6	58,2	68,4	71,1
Washing machine	15,4	31,2	33,4	46,1	58,0	73,7	88,3
Vacuum cleaner	32,4	73,5	78,1	85,3	96,2	98,8	98,9

Source: Bo Diczfalusy, "Energy and income distribution".

Environmental Consideration

Finally we should note that demand for raw materials can be very substantially changed as a result of growing sensitivity to environmental effects or new discoveries concerning health hazards. Thus the Swedish Government recently decided to ban the use of type 2,4,5-T phenoxy acids, even though a symposium convened by the Academy of Sciences not long before had been unable to state categorically that this preparation was harmful to health; the delegates had, if anything, concluded that they were not prepared to make a pronouncement on the subject. Surely, then, there is all the more reason to expect increasingly stringent standards to be applied to the use of metals or chemical compounds which are known to be dangerous?

It is a well known fact that asbestos, cadmium and lead are dangerous substances. Less well known in public awareness is that the majority of metals and chemical substances playing an important part in present-day society entail serious health hazards.

The main impression given by a long list of the dangerous or toxic substances around us is that Sweden entered "the chemical age" long ago, with all the health hazards that this in all probability implies. The increasing attention focussing on the occupational environment has prompted research which is constantly coming up with new findings concerning the risks involved with many of the materials produced today.

It is of course impossible to say how extensively new findings concerning the harmfulness of chemical substances will affect their use. Many substances are quite indispensable. Substitutes may be available for others - e.g. asbestos - at the same or slightly higher costs. It is therefore important for both medical and chemical findings to be transmitted to the general public as quickly as possible, and genuinely discussed. The trade unions have played a central role in this connection, and it should be in the long-term interests of industry to take part in this debate.

All of the factors we have been discussing here are "soft" ones, i.e. they are not readily defined in hard figures, and at the same time opinions can differ widely concerning their importance and/or the direction in which they will develop. It is clear, however, that factors of this kind will have a very important bearing on demand for raw materials in the very long run.

Will Sweden's raw materials supplies be adequate?

Although it is very difficult - and has proved impossible within the RoR study - to say exactly which raw materials will be needed, and in what quantities they will be needed even in fifteen years' time, we can safely say that the flow of raw materials to industry will have to increase if economic growth is to be maintained and if our style of living and our structure of production are not to be drastically changed. Will our national raw materials base be sufficient to sustain this continuing expansion, and can the global scene affect our ability to import the necessary commodities?

Beside the purely physical availability of materials there are at least three more factors on the supply side which might influence the availability of raw materials for a small country like Sweden. These factors are the supply of energy, the concern for the environment, and the overall trade framework which could fundamentally alter the prospects.

Energy and Raw Materials

In this report we are concerned with non-energy raw materials, but these materials are palpably linked with energy. It is clear that a physical energy shortage or gradually rising energy prices must have a very immediate effect on the extraction and processing of raw materials. For example, the profits made by the Swedish aluminium industry during the 1960s and the early 1970s would have disappeared if the cost of energy to the producer had been the same as to ordinary private consumers.

Three factors can be mentioned as being particularly important in a future perspective: increased specific energy consumption, increasing use of materials whose production requires a great deal of energy and greater concern with the importance of energy quality.

The grade and quality of ore at which the extraction of different metalliferous minerals is considered remunerative is constantly declining. Lower grades means that more rock has to be extracted (which costs energy) and that more energy has to be applied to the further processing of the ore in order to concentrate the metallic portion. This suggests that more energy has to be applied to obtain the same final amount of metal. New technology, for example in the form of flotation techniques of benefication, has improved the situation and can be expected to do so still further in future. By this method, impure rock can be separated with a specific energy consumption of about 400-800 MJ per ton or rock, whereas more conventional methods demands 7000-14000 MJ per ton.

Secondly, energy is expected to become more expensive. This will adversely affect areas which are already demanding a great deal of energy today. The paper and pulp industry accounted for 17 per cent of Sweden's total energy consumption in 1971. Iron and steel took 9 per cent, the clay, sand, gravel and stone industry consumed 3.9 per cent and mining consumed 1.4 per cent. Thus sectors directly related to raw materials accounted for roughly one-third of the country's energy consumption. The high energy cost of domestic heating is also connected with energy consumption in the raw material sectors, in that efforts focussing on building materials (insulation, etc.) can greatly reduce the amount of energy required for heating purposes. Rising energy prices will however, not necessarily obliged the raw material industries to reduce their output. This will depend on whether competing countries incur the same rises in energy prices, and it will also depend on the extent to which energy prices can be offset by acceptable increases in the prices of the final products.

A conflict of aims can arise concerning timber as a constructional material and as a raw material for other applications. According to some observers, the raw timber from fast growing energy forests can possibly be put to better use, even from an energy viewpoint, as a material than as direct fuel. (The fuel value, however, can be realized when the material has been turned into waste).

An important aspect of the present-day discussion on energy is concerned with the question of energy quality. In a society which is economical in its use of energy, this will be of importance in the raw material sector. High temperatures and concentrated amounts of energy will be more expensive than they are now, while temperatures below about 250°C will be more easily obtainable, even as a byproduct. This will be an advantage in favour of types of material which can be produced more easily at lower temperatures. New techniques will therefore open up a number of opportunities of more energy-saving processes based on readily accessible mineral raw materials. But it still seems clear that the price of energy will have a vital bearing on future problems concerning raw materials.

Against the background indicated above, we cannot discount the possibility of Swedish raw material industries being faced with considerable problems of adjustment due to changing energy costs.

The Environment and the Supply of Raw Materials

It is a well-known fact in Sweden that environmental considerations can profoundly influence the possibilities of extracting raw materials. (This might seem to be repetition of earlier observations, but environmental considerations are involved both in the question as to which raw materials are in demand - a question which we have already considered - and in the question as to which raw materials are produced - the question referred to here.) Problems connected with the deposition of leeching residues in various mines, and demands for the purification of effluent, need not preclude activities altogether, but they may well result in the scale of activities being smaller, and the costs of manufacturing higher, than would otherwise have been the case.

Forest fertilization is another topical issue where the desire for an increase in raw material output conflicts with environmental interests.

Thus environmental considerations are already imposing limits on society's extraction of raw materials. There is every indication that these limits will become increasingly restrictive if economic growth continues. If, however, there is a more prolonged deterioration in the economic climate, it is to be feared that immediate material interests will get the upper hand. Unfortunately there is a clear tendency nowadays for the level of environmental aspirations to be lowered during periods of recession.

So far we have been considering the external environment as something intrinsically valuable. But there is another potential conflict involved in competition for the use of land. Large-scale mineral extraction and continuing house construction mean the withdrawal of land from agriculture or forestry. All we can say here is that stiffening competition is likely in the future to necessitate more explicit priorities, thereby increasing the importance of general planning in this area - for example, in the form of national physical planning.

Imports of Raw Materials

Sweden's ability to import raw materials is essentially dependent on the general conditions applying abroad and on the organization of raw material markets. In Chapter 4 we discussed a number of factors influencing long-term global supply. We came to the rough conclusion that in the long run the industrialized countries would probably get by more or less, even without access to raw materials from the developing countries. This conclusion should on the whole be applicable to Sweden as well. But the moot point here is to what extent Sweden's position as a small neutral country is northern Europe could influence our import opportunities.

Sweden's non-alliance policy means among other things that in situations of international crisis we can hardly expect to be accorded the same treatment as countries belonging to the great pacts. This may mean that we are unable to obtain such strategic commodities as chromium and tungsten. The distribution of strategic commodities between pact members is often determined by negotiations. If Sweden wishes to obtain a certain share of the available raw materials, this will have to be resolved through some form of agreement with (in reality) one party in the presumed conflict. It seems obvious that this can present serious problems of foreign policy. Sweden would be liable to encounter this dilemma if, for example, a shortage of alloying elements were to result from political differences between industrialized and developing countries.

Because of our diminutive stature in the western economy, we have very little chance of influencing the policies of the transnational enterprises. There is reason to suppose that, in the event of a shortage, these enterprises would endeavour first and foremost to cater for demand in their mother countries. There is an agreement between the United States and Canada whereby in the event of a crisis in the nickel market, Canadian enterprises are to deliver primarily to the USA. A number of Swedish enterprises, it is true, have made investments abroad aimed at securing their own supplies of raw materials, but on the whole Sweden is greatly dependent on deliveries from the transnational enterprises of other countries in the raw material sector.

Thirdly, Sweden will be running great risks if the present system of international trade is superseded by bilateral agreements between individual countires. Our negotiating position is very weak compared with that of the United States, the EEC or Japan, in terms of size or market, financial resources and the range of industrial commodities which can be offered in exchange for raw materials. If our ability to entice opposite numbers into bilateral agreements is limited, our means of exerting political and/or economic pressure are practically non-existent.

This is not true of the countries mentioned previously.

There is no reason to dramatize these risks of disrupted supply or discriminatory pricing, but the effects on Sweden would be serious - much more serious than a price rise affecting raw materials in general - and careful consideration should therefore be given to the precautions which Sweden can take to guard against them.

What we have done so far is to point to a number of risk factors which should be included in official analyses of possible raw materials policies. Such policies are discussed in the next chapter.

7 Aims and Means of a Raw Materials Policy

Introduction

In previous chapters we have discussed resources and raw materials in a variety of global and national perspectives. In this concluding chapter, we shall be considering what can be done about the various problems relating to raw materials and what elements could be included in a more active raw materials policy.

This is no easy matter to discuss. First of all we have to consider whether there is any real call for a more active policy. The existence of any such need is not self-evident,least of all in a study of alternative possibilities which is not committed to any fixed terms of reference. Most official surveys and inquiries, after all, are mounted by virtue of a proven or likely need for policy measures. It is our belief, however, that earlier chapters in this book have pointed to a number of uncertainties which in the future may demand forms of intervention that can be broadly included in the context of a raw materials policy.

Aims

Before we proceed to review the various instruments which might be used, we must try to determine the aims to be set for this policy and the way in which those aims can be promoted by various means. Resources and raw materials constitute such a wide sector that the majority of aims conceivable for societal policy can be brought into the picture. It is not really possible to undertake a systematic review of these aims and of the potential conflicts between them. Nor is it clear what means we have at our disposal. Even the relationships between different means, still more those between different means and ends, are obscure if not completely unknown.

What makes things even more complicated is that we cannot say a priori who is going to be responsible for the framing and direction of policy, is it the state, local authorities, the trade unions, business enterprises or private individuals? And are we concerned with raw materials policy for Sweden, for the industrialized countries, or under the aegis of the United Nations?

There is no point in trying to take in all these different levels unless we believe ourselves to have found solutions to a whole number of conflicts of goals which all communities have to contend with. Then again, we have not found an adequate alternative to the traditional analysis of ends and means, even though we realize that the latter can only convey a very narrow picture of the substantial problems that are liable to occur.

The following pages are therefore confined to considering a Swedish state policy on raw materials. This does not mean that an international policy is uninteresting or that the attitude of the individual is of no consequence. We simply believe that others are better qualified than ourselves to discuss what is possible and desirable at the international level. Our main reason for not discussing the 'raw materials policy measures' of local authorities, business enterprises and private individuals, for instance, is that this would involve a degree of complexity beyond the possible bounds of the present study.

Goals

Raw Materials Policy and Priorities

A policy is an array of deliberate measures taken in pursuit of certain goals. Of course, refraining from action is also a kind of policy, assuming that it is prompted by ideas concerning the results which are desirable or acceptable.

Most goals are interlinked one way or another and form hierarchies, with a goal at a lower level most often constituting a means to the achievement of a goal at a higher level. Anybody seeking to list these goals and arrange them into a consistent pattern is liable, sooner or later, to approach the heights at which policy goals - considered as words - become so broad that in practice they are unexceptionable and therefore void of concrete political content. Who, for example, is likely to object to the goals of wellbeing, justice, the preservation of the environment or the achievement and maintenance of happiness?

Nevertheless we should perhaps start at this end in order to identify roughly the 'major goals' to which measures in the raw materials sector can or cannot contribute to any substantial degree. It is possible to be an ardent advocate of certain goals without believing that a raw materials policy is the best way of furthering their achievement. Conversely, it is possible to advocate certain elements of a raw materials policy without in the least espousing the aims which are vaguely associated with them in the normal run of things. Just as opinions in the community vary concerning the way in which one goal is connected with another, different groups and interests attach different degrees of importance to different goals. To make the ensuing discussion more concrete, here is a list of policy goals which are all commonly referred to in public debate. (We may add that they can be found at various levels).

Goal	Measures included in
Full employment Fast economic growth More equal distribution of income Reasonably stable prices Balance of foreign payments Regional balance within the country	economic policy
Peace National security International justice	foreign policy security policy international development policy development assistance policy
Good external environment Good internal environment	environmental policy

Small is beautiful
Decentralization
Gentle living objectives (e.g. moderation, meaning in life, free from stress, non-alienated employment, human contact, etc.) } improved quality of life-policy

In the above list, and in the discussion which now follows, we have included two goals which do not usually feature in Government surveys: smallness of scale and "gentle living objectives". We have done so because these aims play an important part in the Swedish resources debate. They are often grouped together in a goal referred to in the debate as 'improved quality of life'.

The measures included in a raw materials policy can cut across most of these aims and probably many others too. Many of the above aims can be said to be mutually contradictory, which sometimes makes it impossible for all of them to be achieved in their entirety and all at once. In most cases, therefore, a balance has to be struck.

Resource Restrictions

There are two caveats when it comes to finding balanced solutions to these problems Firstly, can the supply of raw materials be so limited as to impose absolute limits, and thereby constitute imperatives precluding the furtherance of the major objectives? Secondly, is the environment so seriously at risk that it creates absolute limits and, consequently, imperatives for the framing, not only of raw materials policy, but also of economic policy in general?

Both questions are in a way concerned with the same thing, namely the physical limits to human activity. In this study we have come to the conclusion that palpable limits in the form of a shortage of the resources needed to cover basic human needs apply today to that half of the earth's population living in poor countries, but that these limits are social and political, not physical, in character, and that they will continue to be the most important limits for the next twenty-five years.

Discussions of physical limits have sometimes led to diametrically opposite conclusions, and those conclusions in turn have led to completely different evaluations of the overriding objectives. Our project group has not been able to reach an unequivocal standpoint on all the major issues touched on here. This is hardly surprising in view of the differences in our basic values, which are eminently relevant in this context. The following argument will therefore be relatively open-ended, though in cases where we have formed clear-cut views we shall of course state them.

We believe that the debate on resources and, above all, discussions concerning a society more directed towards the conservation of natural resources, are concerned not only with consumption in relation to the possible scarcity of resources but also with the 'unnecessary' consumption of resources which has come to symbolize a society and a pattern of production and consumption which many people regard as futile. Waste is misuse, and both natural and human resources can be misused. It has been argued that the closer people are crowded together in modern mega-cities, the more widely separated they become, and that the substitutability of everything and everybody aggravates alienation and loss of identity. Affluence, too, is seen as an appalling misuse when it is geared to poverty and oppression, as is often the case in developing countries.

This criticism of society's development in the modern western world is well known from the ecological movement, in Sweden termed 'the Green Wave', and from the left-wing utterances of the later 1960s. We believe that there is a great deal of truth in it and that modern industrial society displays a short-sighted wastage of natural resources and raw materials. On the other hand we do not believe that a policy aimed at rectifying such shortcomings can be framed primarily with reference to questions concerning raw materials. The raw materials policy which we will be discussing here implies a narrower approach to the subject than some people would perhaps have preferred.

If resources are running out, the resource-conserving society is something worth aiming for, but a society of this kind would not ipso facto present fewer drawbacks than the present-day industrial society. Instead, its technological complexity might be even greater, and it might be even more 'rational', specialized and characterized by an even greater measure of alienation and centralization. Thus even a conserving society can assume various guises, depending on the overriding aims governing its development. Our only reason for pointing this out is that by 'raw materials policy' we mean measures primarily aimed at influencing the supply of raw materials and the use made of them. These measures affect - by promoting or running counter to - other social aims, just as they themselves are of course affected by consideration for other fields than that of raw materials policy.

In the course of debate, demands are often voiced for 'a fundamentally ecological approach'. The meaning of such an approach, however, is not always made clear, and the expression seems to convey different things to different people. Taken to extremes, 'a fundamentally ecological approach' can imply that practically everything the human race devises must sooner or later have a negative, deleterious effect on nature and that, however efficient it may have been from the point of view of man himself, human intelligence is simply something aberrant, essentially alien to a natural system, which otherwise operates with purely unconscious automaticity.

In our opinion, interpretations of this kind are liable to result in politically untenable standpoints, to the effect that the decline and fall of the human race is in the best interests of nature.

In keeping with a 'fundamentally ecological view', but also with reference to the unequal distribution of wealth in the world, demands have also been put forward to a radical transformation of the whole of our economic thinking. General demands of this kind require a considerable number of definitions in order to become concrete. What they really mean is that the social and economic organization characteristic of, for example, Swedish society, should be changed from the bottom up. We do not question the need for Utopias to outline different alternatives with this type of content, but what is often lacking is a description of the way in which desirable changes are to be accomplished. For example, by what political means will economic growth, as a principal aim of the sort of mixed economy existing in Sweden today, be replaced by an 'ecologically responsible policy'? Is this change to be confined to Sweden, or will it include all the developed economies simultaneously? Sweden's freedom of action is in fact severely circumscribed, because her economy is closely bound up with that of the world at large.

We have not ventured any further into this debate because to do so would soon have meant that all the time available for the RoR study had to be devoted to studies of more general political issues, thus obscuring the more immediate problems directly relating to raw materials, our main concern. In the following pages, therefore, we shall try to discuss a number of instruments of raw materials policy, instead of reiterating general and far-reaching reflections. This is a more limited agenda, but it has the compensation of being more concrete. Our discussion will proceed in relation to what we believe to be the underlying aims of present-day policy and also in relation to aims which we believe ought to carry more weight in future politics.

The Dominant Economic Goals and Raw Materials Policy

Growth has been the most distinctive feature of the economic policy and economic development of the industrialized countries during the twentieth century. Generall speaking, the aim of raw materials policy has been to facilitate this growth, i.e. to ensure that raw materials do not impose any constraint on material expansion. The main demands made on raw material supply in this connection have been as follows.

(a) Secure supplies. The consumer country and its industry are anxious to be spared short-term or long-term suspensions or disruptions of supply.

(b) Low and stable prices. The consumers have a vested interest in the 'right' price, i.e. prices which are as low as possible within the framework compatible with continuing long-term supply. The maintenance of low prices for raw materials is perhaps a powerful consideration, but it is hardly an essential condition. So long as the industrialized countries are affected more or less equally by rising raw material prices, the latter are simply one of many factors in the economy for which compensation can be obtained elsewhere.

On the other hand, it is supremely important to each individual industrialized country that its raw materials should not be too expensive in relation to those obtained by other industrialized countries, because otherwise its manufacturing industry would come under a heavy cost disadvantage. As long as trade barriers continue to be dismantled and international trade continues to account for a very large proportion of industrial output, considerations of cost and competition will keep on setting limits to the raw material prices which can be accepted by the individual nation. This in turn limits the ability of the individual nation to develop domestic alternatives of its own, involving the use of more expensive raw materials. Another alternative is to economize on raw materials by taxing them, i.e. making them more expensive. In the latter case, the frontiers have to be closed or else raw material production has to be supported by means of subsidies. (Most of the industrial countries deploy a wide variety of such measures in their agricultural sectors).

The heavy fluctuations of raw material prices play an important part in international negotiations. The adverse effects of unstable prices are primarily felt by the exporting developing countries. The main focus of attention in industrialized countries with stronger economies is on the possible inflationary effects of these fluctuations, though their effects on outward balances can also be considerable. Some consumer countries, but by no means all of them, have shown an interest in stablizing prices.

(c) Freedom to manoeuvre is an objective of companies and states alike. Even a 'secure' supply of raw materials - if it comes from only one or a few countries - can put a nation in a position of dependence which by other criteria is considered unacceptable in both peacetime and war. Business enterprises too have a similar and natural interest in protecting themselves against pressure and unexpected events, and they therefore try if possible to spread their sources of raw materials or to gain control of them. Freedom to manoeuvre, therefore, is an aim which is closely bound up with considerations of foreign policy and industrial strategy.

(d) Environmental consideration. It is perhaps arguable that environmental considerations have only recently become a serious potential restriction on the economic development of modern industrial society. As has been made clear in other parts of this book, we regard environmental demands - primarily manifested as inevitable local or regional environmental damage, e.g. acidification due to sulphuric emissions - as the physical limit which the affluent part of the world is now coming up against. Raw material industries involve several conflicts with environmental interests of different kinds. Examples mentioned previously have concerned the debate on 'rational forestry' (which among other things entails the use of pesticides, artificial fertilizers, large scale soil preparation and clear cuttings), the possible exploitation of mineral deposits in the Swedish mountain range, the exploitation of shale deposits in the provinces of Narke and Skane, and so forth. It is to be expected - without attempting any prophesy of the outcome - that conflicts of this kind will become increasingly common in the future.

(e) Apart from secure supplies and the 'right' prices, both of which can be said to derive from the growth objective, general economic aims such as increased employment, regional economic balances and the international trade balance often play a strikingly important part in raw materials policy. One of the arguments used in favour of the extraction of minerals in the Swedish mountain range is that the province of Norrbotten cannot live by conservation of nature alone. There are several small communities where mining is perhaps the only industry with a chance of becoming competitive, or at least profitable to the national economy. One reason why the external trade balance is often quoted in connection with raw materials is that their prices and volume often fluctuate heavily.

The ideas underlying a much debated major national effort in the steel industry (the Swedish Steelworks 80 project) and discussions concerning a higher level of processing in the Swedish forest industry are matched by deliberations among exporters in other countries. Why export unprocessed raw materials when upgrading activities within one's own borders can boost the economies of backward areas, raise the total level of employment in the country as a whole and improve the national trade balance? Moreover, upgrading at subsequent stages of the manufacturing process can yield greater value added in comparison to energy input, for example, than the first (raw material) stages, which often require very heavy energy inputs. Developing countries have long been asking themselves the same questions. Current economic theory replies that the resources needed for such an expansion of processing activities can perhaps be put to other, more efficient uses in pursuit of the same objective. But discussions in Sweden and other countries have shown that alternatives are not always that easy to find.

Raw Materials Policy and Goals Differing from those Traditionally Dominant

1. Environmental imperatives. What will happen if environmental considerations cease to be a consideration made in pursuit of the economic objectives, and instead become an unequivocal, inescapable and absolute limit to human economic activity? The long-term uncertainties surrounding energy consumption and climatic change, the increasing acidification of precipitation due to the combustion of coal and oil, and the rising number of mercury, asbestos, PCB and other 'discoveries' in our chemicalized society, are sufficient to make this question worth asking. But nobody today can predict with certainty that we will be unsuccessful in our efforts to solve our environmental problems. Our own view is that a great deal can be done to alleviate the negative environmental impacts, but we believe the possibility that global ecological limits - e.g. for carbon dioxide immissions - will ultimately circumscribe raw material extraction and total industrial output is big enough for us to start thinking today about methods and instruments which can make possible a transition to zero growth or - why not? - to 'the economically decending society'. Clearly one of the first things we will have to do is to see whether GNP growth can be given a content which will make it unnecessary to come

A delicate interrelationship exists between a society and its resources base. Shortage of water can threaten an entire civilization. The nomad tribes south of the Sahara used to be prepared for intermittent years of drought. The drilling of wells brought a temporary improvement of the resource base, and people adjusted to the more abundant supply of water. When the drought finally came, the inherited, culturally-conditioned stamina had not survived sufficiently, and disaster ensued. (Picture by Bjorn Henriksson, Stockholm)

up against the limits dictated by possible environmental restrictions at the global level. A larger proportion of services - both public and private - would probably help to produce such an effect. One cannot, however, discount the possibility that measures of this kind will merely decelerate, instead of halting, the approach to these limits, because the proportion of services contained by each new unit of GNP does not of itself tell us anything about the total extraction of resources. The economizing which may be dictated in the future by absolute environmental limits would demand a restriction of the total extraction of resources.

Seen in this perspective, a simple limitation of the resources extracted for consumption in the industrialized countries will not be sufficient. New arrangements for public services demanding heavy resource inputs (e.g. a reorganization of the system of transport services) or the development of new products which make less use of natural resources may, however, provide a certain amount of scope for improvement within the framework of a constant rate of resource extraction. It should not be unreasonable to expect that at least this scope for improvement should be allotted to those members of the world's population who are most in need of it.

2. International equity. The uneven distribution of the world's riches can be illustrated by differences in GNP. GNP can also be taken as a rough yardstick of resource consumption and of the consuming power of the individual country, i.e. the amount of the earth's resources which that country consumes. But at the same time, it can be a measure of a country's creative power, though admittedly such power is often wrongly applied and developed under international conditions which are unequal and unfair and which widen the gaps in the world.

Unfair distribution can be studied from two vantage points which are really mutually complementary. Firstly, one can see how enormous the differences in resource consumption are, and can accordingly infer that the best way of reducing those differences is to cut down consumption where it is greatest and transfer what is thus saved to those who at present consume least.

How is consumption to be brought down? Good intentions notwithstanding, there is a latent danger in comparing 'wastage of resources' in the affluent countries (energy, food, water, timber, minerals, etc.) with the 'shortage of those very same resources in the poor countries. It is easy to draw the conclusion that we must cut down our consumption of one commodity or another so that something will be 'left over' for other people. But economic relationships are not that simple; a product left unconsumed in one country is not automatically placed at the disposal of other countries. Besides, there is little likelihood of a majority of the people of Sweden, for example, reaching consensus as to which goods they are to abstain from. The most immediate course for the individual citizen is to decide to go without the things which he or she finds most dispensable or perhaps dislikes most - non-returnable bottles, motorcycles, paper sheets, steaks, luxury cruisers, or the combinations of commodities making up charter trips to Majorca, or motor traffic in city centres.

It seems unlikely, therefore, that a consensus would be reached concerning which things were best dispensed with in the name of solidarity. A reduction of commodity consumption right across the board, for lack of such a consensus, seems an equally unlikely candidate for consent, in Sweden or universally.

One is therefore led to conclude that there is not much sense in reducing consumption of specific goods in order to alleviate international injustice. In other words, the goal of international equity cannot really be pleaded, for example, as grounds for reducing meat consumption. On the other hand, other goals, such as public health, environmental considerations or national security, may present cogent reasons for cutting down on consumption of particular goods.

What we can and should do instead is to transfer a larger amount of uncommitted purchasing power to poor countries. This can be done by means of collective saving (higher taxation) and increased international development assistance. There is a great difference between these two methods; instead of deciding what is least necessary for ourselves, we let the countries receiving our assistance (purchasing power) decide what things they find necessary for themselves. It is probably more difficult, however, to gain support for a traditional view of this kind than for more dashing comparisons of abundance and distress.

Reapportionment of purchasing power through increased international development assistance is one way of tackling the inequalities of the world. The other would be to try to revise the conditions governing international trade, which tend to distribute the proceeds of that trade in such a way as to broaden existing income gaps. This is what the demands presented by Third World governments at international negotiations for a New International Economic Order (NIEO) are all about.

It is none too easy, however, to analyse the old international order, and it is a good deal more difficult to analyse a new one. In the first place, differences of income within the majority of poor countries are greater, relatively speaking, than the differences between poor countries and rich ones. Internal distribution within each developing country is not discussed in the context of the NIEO, and the developing countries are not disposed to raise this issue at the United Nations, for example. But the fact remains that internal policies are crucial. Insofar as many of the governments demanding a New International Economic Order represent, in their own cuntries, local elites and upper classes who actually do quite well out of the old order of things, there is no reason to suppose that they intend to profit less by the new order or that the new order would reduce inequalities of income within the developing countries.

Of course, when people in the industrialized countries argue that the internal policies of the developing countries are the key factor in an improvement of conditions for the mass of the population, they risk being accused of ducking the issue. Our intention is to point out that the current proposals for a New International Economic Order will hardly do anything to change the fundamental international structure, which comprises an industrialized centre (consisting mainly of the OECD countries) and a heterogeneous collection of peripheral countries whose economies are appendages to that of the central countries, and whose upper classes often have a great many interests in common with the leading strata of the central countries.

This is no reason for opposing the various proposals put forward by the developing countries concerning a New International Economic Order, but it is a reason for analysing the various proposals separately. Many of the demands put forward are concerned with raw materials. They do not envisage less consumption of raw materials by the industrialized countries - even though certain proposals are aimed at increased processing by the developing countries of their own raw materials. On the contrary, the developing countries wish to sell more of their raw materials and at better prices. Methods recommended for sustaining prices include producer co-operation, raw material agreements between producers and consumers, and an indexation with the prices of the finished goods which developing countries import in exchange for their raw materials exports. The industrialized countries have objected to indexation on the grounds that it is barely feasible on purely technical grounds; that it would demand total state control of international trade; and that it would not do the developing countries any good anyway. This is probably true. The industrialized countries have maintained that cartels hardly favour the producing countries taking part in them, at least not in the long run. This criticism has probably been prompted more by a fear of cartels actually affecting raw material prices than by genuine concern for the interests of the raw-material-exporting developing countries.

Raw material agreements as a method of price stabilization have been tested in a number of instances and with varying degrees of success. The industrialized countries have so far not been very interested in agreements of this kind, in international buffert stocks or in efforts to control supply.

From our previous discussion, we may venture to conclude that, as far as raw materials are concerned, the proposed New International Economic Order may not contain very many innovations that are likely to have a vital impact on supplies of raw materials in the industrialized countries, Sweden included. It should therefore be quite 'safe' for Sweden to support proposals concerning raw material agreements to reduce price fluctuations, combined with programmes of income stabilization, but we should not expect this to result in a big step forward in the direction of a New International Economic Order that will fundamentally transform the conditions of international trade.

Developments in every country are determined above all by human labour and ingenuity and by the methods used in organizing these things. Consequently there is a third way of helping to reduce the injustices in the world, and that is to give both economic and political support to countries and movements which have themselves espoused the long-term aim of shaking off economic dependence, and of organizing their own forces to the best of their ability in the interests of the majority of their population. A developing country economy with this principal aim may possibly reduce its exports of raw materials in the long run, partly because its production of raw materials will focus on catering for domestic needs (for example, foodstuffs instead of plantation crops) and also because an emergent national industry will be able to take care of a larger proportion of the country's raw materials.

It is an indisputable fact that, during the present century, both in the years preceding the last war and during the 1950s and 1960s, western great powers used military and political intervention in various parts of the Third World in their endeavours to oppose movements and regimes wishing to transform their economies and their production of raw materials for the service of national rather than foreign interests. Of course, there have also been strategic and ideological backgrounds to such interventions, but insofar as these measures have been successful (which they quite often have been), they have resulted in developing countries continuing to be ruled by regimes with a vested interest in the preservation of the social status quo and the continuation of exports of raw materials.

It is admittedly also true that in cases where socialist regimes have gained power despite the resistance of leading western countries (as in the cases of Algeria, Cuba, Vietnam, Angola and Mocambique), there has been no emphatic change of course away from the traditional exports of raw materials. On the contrary, these countries are also trying to boost their exports, and more often than not, their potential exports are raw materials. But this is the situation today - we do not know what will happen to exports of raw materials when these countries have achieved a higher standard of living - and it is subject to the important proviso that raw material exports in these cases are the result of deliberate national economic planning and state control of foreign trade.

If our assumptions concerning long-term developments are correct, the conditions governing the supply of raw materials in the industrialized countries will to a great extent hinge on the internal struggle for power within individual developing countries. The standpoint adopted by the industrialized countries concerning that struggle will thus, in a manner of speaking, be part and parcel of their raw materials policy. Reduction of the consumption of individual raw materials will then be less important than support for those efforts in the poor countries which are aimed at ultimately achieving economic emancipation from the industrialized countries. In this way, active Swedish support for such efforts can be said to have a bearing on our raw materials policy.

3. 'Soft' living objectives. There is also cause for some discussion of 'soft' living objectives in connection with our consumption of raw materials, above all because discussions of the negative aspects of our acquisite society take as their starting point the most blatant 'symptom', namely mass consumption of goods requiring heavy resource inputs. But the heart of the matter concerns the social and human costs entailed by the efficient industrial society. A society that is directed towards the conservation of resources is therefore often presumed to lead to balance and harmony both between people themselves and between man and nature. If palpable social problems, such as alienation, lack of identity, segregation (not only between races), the materialization of human relationships, and so on, could be solved by cutting down on consumption of gadgets and raw materials, then this would naturally be a very important task of a raw material policy. But negative traits such as those mentioned above have other, more fundamental causes, which are to be looked for in the way in which modern society has organized production and consumption, work and leisure, housing and the family - in short, the way in which it has organized interpersonal relations.

Large-scale, rational and excessively 'efficient' organization, as prompted by considerations of profitability and productivity, is the arch-enemy of the 'soft' living objectives. The production which this organization brings forth and the 'consumption' which it generates are mere symptoms. As long as people continue to demand more and more products and as long as no greater opportunities are provided for self-realization without consumption, these shortcomings will continue to exist. In other words, demand helps to reproduce the system, and the system therefore has to sustain and augment demand. This is exactly what the system does, partly by means of advertising, fashion and an 'industry of discontent and awareness' which exploits people's uncertainty, lack of identity and loneliness in order to raise demand for consumer goods.

The ever widening debate in recent years concerning the reasons for discontent with the patterns of consumption in the most affluent countries might be summarized as follows:

> Consuming more does not make us happier.
> If we were happier, we would probably consume less.
>
> This does not mean to say that:
> We will be happier if we consume less.

Of course, a summary as brief as this cannot do justice to the debate, and the above formula or anti-syllogism is open to two immediate objections. Firstly, the meaning of "we" is unclear. Even in a country as affluent as Sweden, there are large groups of people who do not come up to a reasonable standard of living. They will obviously be happier if they consume more. Thus one cannot disregard the distribution of income. Secondly, our argument refers to consumption of goods. Few would claim that even the most affluent societies have a perfectly complete structure of social services. There are pressing demands in Sweden for improvements to medical care, dental care and child care.

'The other society', often referred to in debate as the resource-saving society will, in addition to reducing consumption of goods, probably have to devise forms production and social organization which are different from those with which we are familiar if it is to attain the softer objectives of life. Probably "a radical transformation of the whole of our economic thinking" will also be called for - a transformation so radical that its implications extend far beyond the question of Sweden's supplies of raw materials. In other words, we do not believe that the instruments properly termed instruments of raw materials policy are the best means of promoting the 'soft living objectives'. On the other hand, the instruments of raw materials policy should be designed so as not to impede the attainment of those objectives. In our efforts to meet our supply requirements, therefore, we should try to avoid getting into a situation where the decisions we have taken lead us to resist (or to increase our previous resistance towards) demands from the poorer countries for reasonable changes. The shaping of a raw materials policy should also be made to incorporate similar consideration for the environment and for the quality of life in a broad sense

Perhaps our conclusion can be phrased as follows. Cutting down on consumption of raw materials is not the most immediate measure by which we can pursue the aims of international solidarity, environmental conservation or a higher quality of life. On the other hand the shaping of a raw materials policy can be important when it comes to supporting other, more wide-ranging measures taken in pursuit of these ends.

Instruments of raw materials policy

Although discussions concerning the need for an integrated policy on resources and raw materials are a relatively new phenomenon, state policy with a bearing on individual raw material sectors or particular problems concerning natural resources has existed for a long time. This is immediately confirmed by a quick review of the long line of official Swedish Commissions concerning agriculture, forestry and mining. On the other hand, the post-war period can hardly be said to have witnessed a raw materials policy in the sense of an integral policy concerning supply. Policy on natural resources between 1945 and the 1970s - as manifested by official surveys, Government Bills, parliamentary debates, etc. - has been dominated by demands for efficiency and structural change in primary sectors and industry; considerations of conservation of nature and environment protection have taken a back seat and have featured principally as measures of adjustment and restraint. Even then, these measures have mostly been applied against fairly marginal but conspicuous abuses or patently negative consequences of a particular type of consumption or production. The recommendations concerning more general consideration for nature and the environment which have been issued in connection with Governmental Commissions have often been so vague as to be devoid of any practical significance. Lack of knowledge and the fragmentation of administrative responsibility between a large number of authorities have probably been among the main factors contributing to this state of affairs. (A survey of water and sewerage problems in Sweden soon after the Second World War revealed that no less than twenty national authorities and four Ministries were responsible for such matters.)

In comprehending and formulating problems and drawing conclusions regarding the changes which are appropriate or necessary, decision makers have been thrown back on existing structures. Some problems are found to be interesting, others uninteresting. Very often there has had to be a pronounced tension, and 'achievement gap', in order for change to occur. These existing structures and ideas have tended to have a conservative influence.

One example of this type of rigidity is the inevitable division of public expenditure into different allocations. The discussion surrounding ownership of land and responsibility for preserving its natural endowments is a typical example of the effects of budgetary compartmentalization. Proposals to the effects that the rules of dilapidation should focus on nature conservancy and care of the landscape, as well as on agricultural efficiency, were rejected during the 1950s on the grounds that the costs of nature conservancy should not be borne by agriculture (and consequently via food prices). In other words, nature conservancy was to be provided for in nature conservancy policy and not via agricultural policy.

During the 1970s, some progress has been made towards a more integrated approach to the management of natural resources, not least as a result of national planning of the physical space. At all events, there has been a growing realization of the need for the economy and production as a whole to be viewed in the perspective of environmental policy, and there has also been a growing realization of the inadequac of fragmenting one's attention between individual, subsidiary problems.

Are we making headway in the same direction where supplies of raw materials are concerned? All the mechanisms which have delayed the emergence of an integrated approach to the environment are probably active in the raw materials sector as well. In addition, of course, one can ask whether it is possible for all raw materials to be dealt with in a single raw material persepctive. Many of the instruments discussed below embody principles which are not new, and the majority of them are already being employed today by governments and parliaments. This applies to the energy sector or energy raw materials which in Sweden, for example, are a subject of taxation, intensive research, opinion mobilization (under the aegis of the Energy Conservation Committee), extensive state prospecting, etc. The following discussion will be concerned to some extent with the question of the applicability of this repertoire of instruments to other raw materials besides energy, if there are fears of future shortages.

Measures to Reduce the Use of Scarce Raw Materials

Resource-saving production methods

Earlier in this book, we have seen examples of changes in the material intensity of production processes. Technological progress in constantly feeding this tendency One example of a possible technique which is more economical in its use of materials can be taken from metallurgy. With the methods now used for turning metals into structural parts by means of welding, turning, etc., a good deal more raw material has to be put into the manufacturing process than comes out of it in the form of products. It has been estimated that 70 per cent of the raw material used in metal manufacturing is machined away. Most of the waste resulting from these processes (swarf, etc.) is salvaged and remelted, but this still leaves a considerable net loss, partly in the form of non-recoverable waste, and also partly in the form of transport costs and the need for further processing. Thus processing and joining techniques of a new kind can save resources.

A great deal of research is already being done into substitutes, i.e. alternative raw materials. Examples include ceramics, superplastic alloys, etc.

Substitution research has always formed part of overall industrial strategy; it is prudent to have alternatives lined up in case the raw materials now being used should become more expensive. Substitution research can also be described as a natural reflection of the efforts made by companies to keep to the fore, technologically speaking. Growing uncertainity regarding the availability of raw materials appears to have provided a new and powerful incentive for the search for substitutes

But substitution can mean other things besides the replacement of traditional raw material with a new one. One can, for example, replace one system with another (public transport can replace private motoring, for example) and this is naturally bound to affect consumption of raw materials.

These tendencies are already operating without state support. A more active governmental raw materials policy would be aimed primarily at stimulating them and guiding them into channels which are judged to be beneficial in the long run. The most powerful argument for doing this would be that the state commands an overall perspective which business enterprises are neither capable nor desirous of achieving. One of the first measures heralding a more systematic policy on raw materials should therefore be an acceptance by public authorities of the task of closely and continuously monitoring the raw material sector, a function which cannot be performed by ad hoc Governmental Commissions. In this connection, there is reason to consider whether co-ordination under the auspices of the state can promote a technology less demanding of scarce resources. A third method of promoting a resource-saving technology is that of direct support for research and investments. National and local government purchasing can also be used as a means to this end.

The above argument is based on the premise that it is desirable to save raw materials, either because they are expensive or can become so, or else because their future physical availability is uncertain. Resource-saving technology is also aimed at securing national supplies, at restraining prices and at increasing our freedom to manoeuvre. A policy of this kind, however, can conflict with other objectives.

First and foremost, it may come into conflict with the aim of international justice, at least in the short term. If raw material intake is successfully limited by all rich countries, this will (among other things) reduce dependence on raw material imports from the poorer countries. Many raw materials - alloying elements, coffee, raw rubber, etc. - come from the developing countries, and so these countries would suffer a reduction of their export earnings. The growing payments deficit incurred by the developing countries after the Korean boom and until the end of the 1960s - their raw material exports having declined in value in relation to their imports of industrial products - is partly connected with technical advances in the extraction and processing of raw materials and in the development of a number of substitutes. At several international conferences, the developing countries have in fact voiced their opposition to investments and research focussing on synthetic products, e.g. artifical fibres, instead of natural products. It is admittedly arguable, in a more long-term and scientific perspective, that expanding knowledge broadens the opportunities available to all mankind, and that a genuine New International Economic Order could conceivably result in technical knowledge being transferred in such a way that everybody shared in its benefits. Until this happens, it is to be expected that the countries or interests possessing a fund of knowledge will exploit it to the greatest possible extent. Thus technical know-how will be - and is - an element in the distribution of power.

Another possible conflict of aims could result from the increased destruction of the environment entailed by a technology giving one-sided priority to the conservation of raw materials. One such technology is that which gives more efficient utilization by virtue of increased scale. In many cases, a technology of this kind is resource-saving in the narrow sense but, at the same time, harmful to the environment. There is no intrinsic reason why new technology in the raw materials sector should have graver environmental consequences than has traditionally been the case, but there is a clear tendency towards increased impact on the external environment and towards large-scale interference. There is no likelihood of the tendency towards large-scale, hazardous technology being reversed until the environment actually becomes the governing factor in research policy.

Declaration of raw material use

In a Bill put before the 1975/76 session of the Swedish Riksdag, it was proposed that companies should be placed under an obligation to declare their consumption of energy and raw materials in the production of consumer goods. First of all, a

Government Commission was to be appointed to see if this was a feasible arrangement. The grounds given for these proposals were: "Our natural resources are finite and we must economize on energy. .. Solidarity with the peoples of the world therefore demands of this highly industrialized country that we shoulder our responsibility and carry out a long-term transformation of our pattern of living".

The idea of a declaration of energy and raw material content, giving consumers a better idea of the resources which they are actually consuming, has been adopted by other protagonists, often with the aim of providing documentation on which to base criticism and debate concerning the product policies of business enterprises.

The most interesting proposals seem to be those providing for the compilation and publication by companies of specifications showing, among other things, raw material consumption during the service life of the product, including possibilities of recycling. Evidently, this principle is already being practised in shipbuilding, for the scrap value of a ship constitutes quite a significant portion of its selling price.

Propaganda for a resource-conservation society

An active campaign is today being conducted in Sweden in the energy sector by the Energy Conservation Committee. Although the effect of this campaign have probably been rather limited, it is an interesting example of the reaction of the government and parliament to the perceived threat of future shortages, There are no essential differences between energy and other raw materials in this respect. During the last war, propaganda for economizing in various fields was common practice in Sweden, as in most countries.

One of the fundamental problems concerning propaganda is that is is often hard to convince people of the likelihood of a future shortage. On the other hand, once the shortage has developed, propaganda may be superfluous. Propaganda therefore seems most productive when combined with measures of other kinds.

Raw material taxation

In the debate on resources, it has been proposed that raw materials should be taxed in order to reduce consumption and thus to conserve resources for the benefit of future generations. In the United Kingdom, for example, the Conservation Society has proposed a special tax on North Sea oil to slow down the extraction rate. Professor Hayashi of the Japanese Institute of Future Technology has recommended a global tax on resource consumption, the proceeds of which should primarily go the developing countries. We may also mention the RIO report,[1] which proposes an international tax on the profits of oil extraction (70%) and on the production value of minerals (50%). The main purpose of this taxation would be to redistribute the proceeds of mineral extraction for the benefit of the developing countries.

Taxation of natural resources is an inviting expedient when long-term supplies are a topic of concern.

We have already mentioned the taxation of energy, but the main purpose of this seems hitherto primarily have been to increase national government revenue rather than to limit total energy use. In the proposals put forward by President Carter in April 1977, however, energy taxation has been given the task of helping to restrain the growth of demand.

(1) See above page 102

It is easy to draw a parallel with the discussion concerning environmental taxes, the idea of which is to influence economic decisions by levying a charge on the utilization of a resource which, in the eyes of the community at large, is both limited and valuable, i.e. a clean natural environment. Within ten or fifteen years' time, economists will probably be campaigning as actively for intervention in the pricing of natural resources as they are today concerning the environmental sector.

It is not really possible to discuss raw material taxation in general terms. If we start with national taxation, it is obvious that a tax (or a customs duty if the commodity is exclusively imported) raises the price to the user. If the user's price sensitivity is low, consumption of the commodity will not be greatly reduced. Price sensitivity is very much dependent on time horizons. In the short-term, demand is likely to be fairly rigid, but in the long term companies will change to other raw materials and consumers to other products.

Another question is the stage at which a raw material is to be taxed. The advantage of taxation at an early stage - i.e. 'at source' - is that the effects are then shared by the rest of the economic system, so that in principle the least 'productive' or most dispensable uses are hit hardest. However, one cannot be certain that the process of economic adjustment will operate in a manner conducive to the achievement of other important aims. For example, the most 'productive' use may result in the raw material being entirely consumed by its purchaser, while other, less 'productive' uses may be compatible with recycling. In a situation of this kind, a differentiated tax could perhaps be employed so as to make recycled raw materials cheaper than their virgin counterpart.

Reduced consumption of scarce raw materials should constitute a step in the right direction in response to possible ecological imperatives. Assuming correct prior identification of the raw materials destined to run short, it can also facilitate the adjustment process within business enterprises and consumption, but it will run counter to other objectives.

The Swedish Government is not necessarily at liberty to raise its tariffs within the present system of international trade, because we have certain obligations under the General Agreement on Tariffs and Trade (GATT) and under our agreements with EFTA and the EEC. Whether the tax on raw materials is levied at our frontiers or not, the countries exporting raw materials will probably react strongly because the tax reduces their export earnings. It will do so because the volume of demand will contract, and also because (in the majority of cases) the imposition of a tax will reduce the price per kilo obtained by the exporter. It is easy to imagine how Brazil would react if the Swedish tariff on coffee were to be reintroduced, or how Malaysia would react if a tax or duty were imposed on natural rubber. There would be a similar reaction from Swedish producers of raw materials - fishermen, for example - if demand for their products were restrained by the imposition of charges. In the short-term, at least, the goal of international justice as discussed at the United Nations would not be brought any closer, because the developing countries would suffer a decline in their export revenues.

Consumers of raw materials will also be affected. If Sweden unilaterally introduces a raw materials tax, her international competitive strength will probably decline, at least in the short run. True, the competitive strength of business enterprise is not wholly determined by the prices of raw materials; it is no less dependent on educational policy, taxation policy, regional policy and so forth. But a tax on raw materials would have the net effect of shifting the balance of competition. This is a matter which will have to be closely analysed should the question of raw material taxation arise. It is equally obvious that a close investigation will have to be made of the distributional effects of a tax on raw materials, because a tax of this kind affects the relative prices of raw materials, capital and labour.

Some of the conflicts of goals referred to above may prove to be of minor consequence, particularly in the long-term. If the 'aim' of preserving the competitive strength of industry is an overriding one, a raw material tax could conceivably be offset, for example, by a reduction in company taxation. If Sweden were to be ahead of the rest of the world in realizing that a particular raw material was destined to run short, this would ultimately bring an enhancement of competitive strength because industry would then have changed gear at an earlier stage. When the shortage came, i.e. when prices rose, the tax on raw materials could be reduced again and company taxation once more increased.

It may be objected that fiscal manipulations of this kind demand a greater knowledge of the future than we possess today, and that they also demand a more flexible taxation policy than has been usual hitherto. Neither method, however, can a priori be discounted as completely impossible.

Some of the conflicts of goals mentioned above would be eliminated by an international tax on raw materials. A tax of this kind could, for example, imply - at least - theoretically - that all consumers incurred the same increase in overheads. An arrangement of this kind would also be more conducive to international justice because, for example, the bulk of the revenue could be transferred to the poorer countries.

It should already be clear that this idea implies a considerable change in the existing structural framework of international economic corporation. If the state of competition is to remain unaffected, it will obviously be necessary for individual countries to accept an international tax applicable to domestic as well as foreign transactions. Thus producers of raw materials - e.g. Zaire (copper), Australia (bauxite) and Sweden (pulp/paper) - would be forced to introduce a national tax aimed at limiting demand for important export commodities and at preventing the country concerned from deriving 'dishonest' benefit from its own natural assets.

It is possible that other international tax agreements may be more realistic than arrangements of a global nature. For example, a number of countries may arrive at a common assessment of the future raw materials situation, and may accordingly introduce national taxes at uniform rates. It is true that the competitive strength of outside countries will be increased as a result, but the participant countries may be prepared to accept this.

Summing up, taxes on raw materials are of interest as a possible means of establishing a system of resource-saving industrial production. In the majority of cases (and assuming that global shortages of individual raw materials develop gradually and not abruptly), taxation or other price-raising measures should be preferable to physical control (rationing). In this brief review, we have only been able to indicate some of the problems accompanying national or international taxes on raw materials. Taxation, however, seems to be such an effective means of restraining consumption and transferring purchasing power internationally as to merit a closer analysis than it has so far been subjected to.

Rationing

Decisions to prohibit the use of raw materials or to impose physical limits on their use are not as uncommon as might be supposed. Environmental considerations have long imposed clear limits concerning toxic content, for example. An expanding corpus of environmental legislation has developed in recent years. The banning of DDT and certain phenoxy acids, and the debate e.g. in Sweden on forest fertilization also show that decisions of this kind may very well have substantial effects on the production of raw materials. There have been many cases where regulations on permitted upper limits for emission of certain substances have in effect rationed the use of individual raw materials.

There are other considerations besides those of an environmental nature which can prompt physical control of the consumption of raw materials. For example, it may be found desirable for reasons of regional or distribution policy that an incipient shortage of raw materials should not be distributed throughout the economy solely through the agency of prices. Furthermore, the question of controls may arise if Government and Parliament do not believe that the market accurately reflects the long-term availability of one or more raw materials.

If Government and Parliament decide to ration raw materials, this will mean specific quantities being allotted to individual producers of semi-manufactured or finished goods. (This happened in Sweden, as in most other countries, during the Second World War). In such a situation, one is compelled to decide which forms of consumption should be given priority and which should be counteracted. As we have seen, however, rationing is not unique in this respect. Effects on different income brackets are of course taken into consideration in the framing of taxation policy today, but they are also considered, for example, when fixing milk prices. The advantage of direct controls is that the intended reduction of consumption is achieved almost immediately. Their main disadvantage is that they introduce new rigidities in the economic system, which can lead to bottlenecks, with shortages of a raw material in a sector needing it for production purposes and gluts in other sectors where the raw material in question may be dispensable. One does not have to go very far afield to discover instances of the difficulties involved in getting a planned economy to run smoothly. These difficulties have to be weighed against the nature of the problems which physical controls are meant to solve. Perhaps we will not have a free choice in the long run?

Measures to Increase the Supply of Raw Materials

Geological surveying and prospecting could broaden Sweden's resource base, but isolated prospecting operations are not really the most efficient way of achieving this. It is arguable that prospecting can only result in a distinct augmentation of reserves when geological knowledge is employed in close conjunction with a knowledge of new technology in the rest of the processing chain for the mineral concerned.

Increased recycling can mean a greater supply than would be available if only virgin raw materials were used. (The problems connceted with recycling have been treated in one of the sub-studies forming part of this project). The conclusion drawn in that sub-study for Sweden is that no major gains in terms of raw materials are to be anticipated within the near future. The most immediate possibility appears to be the recycling of household waste paper, but this can today only make a limited contribution to Sweden's raw material base - possibly in the region of 300,000 tons in 1974, of a total output of 5.5 million tons. Theoretically, central waste processing facilities could result in the recycling of 700,000 or 800,000 tons.

As regards plastic materials the main opportunities for recycling are probably to be found in the processing industries and commerce. Systematic retrieval of plastic from household waste seems to present problems, and the tendency here is apparently more in favour of recovering the energy from the plastic content of waste by means of incineration or pyrolysis.

Similarly, the main opportunity for recycling metals is probably to be found in a more careful sorting of household and industrial waste, but we should not expect to gain any very substantial quantities in this way. The difference between the steel estimated to have been scrapped and the amount recovered is about 500,000 tons. Raw steel output in 1974 was almost 6 million tons. The data in table 18, referring to conditions in the United States, will furnish some guidance where other metals are concerned.

Table 18. Current and possible recycle percentages in the USA.

	Iron and Steel	Aluminium	Copper	Lead	Zinc
Present	47,3	17,3	44,0	43,9	16.5
Possible	48,9	42,6	54,4	47,2	32,6

Source: F E Banks, 'Natural resource availability some economic aspects', Resources Policy March 1977, p.2. The data given are originally from L Fischerman and H Landsberg, 'Adequacy of Non-Fuel Minerals and Forest Reserves', in US Commission on Population Growth and the American Future, Research Report, Vol.III, Washington DC, 1972.

Even in a small country like Sweden, a policy aimed at increased recycling has obvious international ramifications. These are not confined to the existing ban on exports of iron and steel scrap. If a genuine recycling of materials is to be achieved, this will require not only international trade in recovered materials, but also international agreements governing the product cycle. One of the greatest difficulties where the recovery of metals is concerned is that most often the scrap metal contains a mixture of elements, due to surface coatings or due to their incorporation in alloys or in composite materials. A small country can hardly make unilateral decisions concerning product specifications to facilitate recycling, without running the risk of losing ground in its markets. This problem resembles the problems following the introduction of new national environmental standards. Where recycling is concerned, however, international agreements are probably a more pressing requirement, because the country producing and exporting the goods does not necessarily stand to profit by increased recycling.

Even within the individual country, however, there are a number of measures that can be taken to increase recycling. The American study, "Material Needs and the

Comment

The calculations of 'possible recycle percentages' are based on certain assumptions on the growth rate of industrial activity in the future, on the fraction of the material recoverable from final goods and the durability of products. If these assumptions change then the level of 'possible recycle percentage' changes as well. The level of 'possibility' is given in a situation where the recycling is assumed to have bben systematically applied. Due to the relatively small amount of recycling that has taken place in the past, there is an amount of potentially recyclable material available today as to make it possible (for a limited time span) literally to run the US economy on secondary materials (thus putting 'short term possible recycle percentage' to 100%!). What the figures convey is the understanding that a long term recycle percentage of the total input needed - given a certain expansion of industry, fraction of the materials recoverable from final goods and durability of goods - is limited.

Environment Today and Tomorrow", presents nine recommendations concerning federal action, including the following:

- Subsidized credits to private enterprise for the extraction of raw materials from municipal waste.
- Grants to municipalities to help finance increased waste recovery.
- Tax concessions to steel mills if, for example, they use more recycled raw materials.
- Tax concessions to new factories based on recycled raw materials.
- Procurement for state activities should favour recycled materials.

Our own special study on recycling raises the question of whether a national recycling policy is not also called for in Sweden. A policy of this kind could mean, for example, that state purchases centred more on products based on recycled raw materials, especially paper. In the same report, it is asked whether local municipalities ought really to guarantee the profitability of recycling or whether the state should be more directly involved in commercial activities of this kind.

Measures designed to increase recycling can thus range from research through subsidies of various kinds to direct state controls. One is therefore ill-advised to embark on a discussion of conflicting goals without first defining the policy instruments referred to.

Generally speaking, the most important immediate goal of increased recycling seems to be an environmental one. Even where fairly innocuous wastes, such as household refuse, is concerned, recycling is an important means of reducing the volume that is desposed of in various ways.

Recycling also makes it easier to safeguard supplies, because it reduces the quantities of raw materials that have to be imported for each stage of production. On the other hand, it is doubtful whether increased recycling will reduce the prices of raw materials, at least in the short run.

In the course of debate, increased recycling is often related to the 'soft' living objectives. It is presumed that, in a society which is resource-saving and which is based on small-scale operations and self-sufficiency, all products will be recycled. But an industrial society recycling most of its waste products would probably require large-scale waste processing facilities, and a system of rules covering not only product design , but also the reconversion of products into raw materials. Exactly how 'soft' or gentle a society of this kind would be is a debatable point.

Measures to Secure Supplies

Stockpiles

Sweden has emergency stocks intended for use when the country is at war or when there is "war or danger or war in the immediate vicinity of the realm". The goods thus stockpiled were valued in 1975 at about one third of a billion dollars. There are also peacetime contingency stocks, intended for use when Sweden's supplies of important commodities are cut off or impeded without an outbreak or threat of war. These stocks mainly comprise raw materials and certain semi-manufactured products, and their purpose is to maintain production in Sweden.

Several western countries have national stocks of this kind. The strategic stockpiles maintained by the United States are by far the greatest, and France and

Japan are now in the process of building up contingency stockpiles. The total value of important raw materials stockpiled by these three countries probably exceeds 10,000 million dollars.

The stockpiling of raw materials is also being discussed at the international level. IEA, the International Energy Agency, has decided on a minimum oil storage level to assure member countries of supplies in the event of an oil crisis. These stocks - for which a target has been set corresponding to 90 days' normal consumption in 1980 - are maintained by the various member countries, but in emergencies they will be utilized according to rules agreed on previously. There is no question, however, of any pooling of resources.

The UN Food Conference in 1974 discussed wheat stocks of two kinds: one which could be distributed at short notice to cope with specific emergencies, and another to provide the developing countries with a safeguard against bad harvests - a system of international food security. Within the framework of this system, the various countries were to undertake to co-ordinate their national storage policies at the international level. So far, neither of these initiatives has been realized to the extent that was anticipated, although substantial progress has been achieved in 1977-1978.

Finally, we should note that UNCTAD (UN Conference on Trade and Development) has in recent years been actively discussing an integrated raw materials program (bananas, bauxite, cocoa, coffee, copper, cotton, hard fibres, iron ore, jute, manganese, meat, phosphate, rubber, sugar, tea, timber, tin and vegetable oils). The idea is for arrangements to be made which will help to achieve more stable and, in many cases, higher prices for these commodities. Particular priority has been given to ten of them, which are classified as storable, and concerning which efforts will be made to conclude raw material agreements providing for stocks to be built up when prices are low and sold off when they are high - buffer stocks, in other words. Despite the demands of the developing countries, no agreement has been reached concerning the stabilization of prices, at a level where the purchasing power of the developing countires' exports will be maintained or increased.

Clearly, the different kinds of stocks mentioned above serve various purposes. The Swedish stocks are designed to protect Sweden's economy either in the event of actual or imminent military conflicts, or in the event of the market being disrupted due to, for example, producer countries restricting supplies. The main purpose of stocks under the aegis of the United Nations is presumably to stabilize prices and to promote the development of the developing countries. Stocks built up by consumer countries may be presumed to have quite a different purpose, namely to prevent or impede actions on the part of the producer countries aimed at limiting supply and/or raising prices - or at least to alleviate the effects of such actions. In certain cases, the aims are entirely related to foreign policy. One example is given by the US official policy in the oil sector. At the 1975 ministerial meeting of the IEA - the International Energy Agency - Professor Henry Kissinger, at that time US Secretary of State for Foreign Affairs, made a statement on the motives for building up a system of oil stocks:

"At the Washington Energy Conference, we recognized that only collective action could reduce our excessive dependence on imported oil and restore to our governments mastery over our own economies and foreign policies. Separately we could never create conditions for lower oil prices. Nor could any one of us, except at exorbitant cost, defend against a new embargo. Our security, our economic growth, our role in the world were at risk".

But the various kinds of stocks differ not only as regards purposes but also as to their economic and political consequences. Sweden is unlikely to incur any criticism for building up emergency stocks; on the contrary, this undoubtedly improves the credibility of our policy of neutrality in the eyes of the rest of the world. Swedish emergency stocks can hardly become a crucial factor in the world economy.

Sweden's peacetime contingency stocks cannot be presumed to have a very marked effect on markets, or the external political situation either. The same cannot be said of Sweden's participation in consumer clubs of various kinds, where we are actually linked to a small group of industrialized nations. The fact that this can confront Sweden with awkward decisions in the field of foreign policy was reflected by, for example, the special proviso that accompanied Sweden's accession to the IEA. If there is an increase in the number of producers' associations, Sweden may be faced more and more often by the conflict between the goals of secure supplies and the external trade balance and, on the other hand, freedom of action, consistent foreign policy and international justices.

Sweden has supported proposals within UNCTAD concerning what is termed an integrated programme for raw materials. Negotiations concerning a "common fund" to finance stocks of this kind are still in progress, but the original timetable has had to be abandoned, perhaps partly because the raw materials markets have been hit by the recession in the industrialized countries. Producer stocks of many raw materials have grown - cf. iron ore, copper and wheat. The consumer countries are not so worried about supplies of raw materials as they were in 1974. Another reason is the existence of a general aversion on the part of various administrations and major commercial interests in the raw materials sector to interference with the structure of the market and this is delaying and impeding practical investigations of the proposals that have been put forward. Nor can one discount the possibility that the developing countries themselves may have become a little more sceptical, even though they are continuing, to press the issue, for other reasons. Then again, countries exerting paramount influence on trade in a particular raw material do not want (i.e. developing) countries to be given the opportunity of indirectly influencing this trade. Another point is that international stocks can make it more difficult for the producer countries to take policy measures. If a country wished to withdraw from a raw materials agreement, in order to limit supply through a producer cartel, the ensuing price increases could presumably be counteracted by sales from the buffer stock. At the time of writing, several countries (including the United States) have joined in positive statements concerning the co-ordinated finance of an international buffer stock system. It remains to be seen, however, whether such a system will actually materialize and, if so, what ends it is intended to serve.

From Sweden's viewpoint, an international stock means far less security of supply than a national one. At the same time, though, raw material agreements, whether they include stocks or not, can be important to industrialized consumer countries. For one thing, as we have already seen, more stable prices can be an important factor in maintaining the level of investments in the raw materials sector and thus an important factor in the general context of inflation. Raw material agreements may further include provisions concerning non-discrimination as regards deliveries. As a rule, producers also undertake not to limit output except when the agreement makes provision to this effect. Finally, raw material agreements have included provisions guaranteeing deliveries, i.e. an undertaking to deliver certain quantities, perhaps at predetermined prices.

Bilateral or multinational agreements

At times when the international market in raw materials is subject to heavy price fluctuations and great uncertainty regarding future supplies, each importing country tends to safeguard its own interests by means of long-term agreements concerning purchases of raw materials, often combined with undertakings of other kinds (such as development assistance, credits for the extraction of raw materials, and industrial co-operation with the exporting country). The seventh French five-year plan includes a four-point raw materials programme, the final point of which reads: "Co-operation helping the developing countries to prospect for, extract and process their minerals will at the same time create favourable conditions for France in those countries".[1] In the United States, the AFL/CIO has taken a special interest in bilateral agreements on raw materials.

Agreements concerning economic co-operation can cover a wide field ranging from the general exchange of information to economic integration, as represented by the EEC. Consequently there are various levels of aspiration for such agreements. There are certain elements, however, which are nearly always present in discussions of bilaterial agreements concerning raw materials, namely long-term contracts for the delivery of raw materials, usually involving some sort of quid pro quo from the purchasing country, e.g. technical or financial assistance, guaranteed prices and so on. The purpose of this type of agreement from the viewpoint of the consumer country is to safeguard supplies of a particular raw material and perhaps, at the same time, to enlarge the export market for its industrial products.

What reasons can be given in favour of more systematic efforts to conclude long-term co-operation agreements covering supplies of raw materials? One possible premise is that developments during the past few decades have led to increasingly centralised control of economic affairs and foreign trade in several of the developing countries. Another, no less plausible, would seem to be that the international trade system of today affords rather limited safeguards for deliveries to small countries in the event of shortages. If more emphasis were to be placed on bilateral agreements, the principal objective would presumably be to safeguard supplies and full employment in Sweden. On the other hand, Sweden's prospects of purchasing raw materials at the lowest possible prices would be fairly limited, and our freedom to manoeuvre would be circumscribed in several essential respects.

When discussing bilaterial agreements it should be made clear that the establishment of trade in raw materials on a bilateral basis is a hazardous undertaking for a small and highly industrialized country. As such a country, Sweden is dependent on a number of international agreements based on free world trade on (theoretically) equal terms. Special tariff concessions to individual countries for individual commodities would, for example, be contrary to GATT. Previous Swedish arguments with developing countries have been based on the assertion that no such infringements would occur. If they did occur, Sweden would probably incur sanctions of various kinds. Things will not turn out any better, however, if through our efforts to achieve bilaterial agreements, we contribute towards a more general tendency towards bilateralization whose extreme consequences will certainly not be favourable to small countries. Unfortunately, the converse also applies: if Sweden adheres longer than other countries to the principles of free trade, we are liable to be too late in coming to the conference table.

Even without a general tendency towards bilateralization of world trade in general and in raw materials, a number of developing countries may come to display a growing interest in more extensive co-operation with Sweden. To many developing countries, Sweden seems a more attractive partner in co-operation than the big industrialized nations or their transnational enterprises, particularly where the exploitation of raw material assets is concerned. One reason for this is our policy twoards the developing countries, and another is our smallness and the improbability of our being able to exert pressure on our partners.

(1) The RoR study "Politics and raw materials in an international perspective", 1977. A summary of the report is given in Appendix 3.

It may also prove advantageous from Sweden's viewpoint for development co-operation to expand into co-operation concerning raw materials and industry. It is sufficient to observe that four countries now receiving Swedish assistance - Vietnam, Cuba, Angola and Mocambique - have abundant mineral deposits. Swedish co-operation with these countries could mean, for example, long-term agreements for the delivery of minerals at world market prices in return for measures by Sweden - credits among other things - to develop mineral extraction or processing industry in the countries concerned.

One self-evident prerequisite for such an agreement is expressed interest on the part of the developing country concerned. Another, and no less self-evident, is that there must be a long-term Swedish economic interest at stake. This type of co-operation should be clearly distinguished from development assistance, at least from the Swedish end. (To the developing countries, the difference between assistance and commercial transactions is not always so interesting and is difficult to maintain in practice vis a vis the majority of industrialized countries.)

The first reason for this is that there are substantial differences between the aims of development assistance policy and the aims of such raw material agreements. Development in the developing countries on their own terms, and safeguards for Sweden's supplies, are not always compatible. If stated as parallel goods, they might generate uncertainty concerning the means to be employed, and also concerning the way in which effects are to be judged. Swedish development assistance is then liable to mean subsidization of Sweden's imports or raw materials, rather than a contribution towards the development of the Third World countries.

Moreover, agreements of this kind must be based on a genuine interest on the part of Swedish industry, i.e. on an essentially profitable transaction. To present this as part and parcel of development assistance policy would be to base relations with developing countries on wrong premises. But the most important reason is that Sweden would lose if development assistance were to go straight to Swedish industry. Assuming that free trade continues, the sole effect of such a policy would be to reduce Sweden's competitive strength. Exports would be switched to markets where profits were guaranteed by the state. Experience also indicates that tied assistance, instead of stimulating efficiency and enterpreneurship, tends to shove industry in to sheltered "hunting grounds" of different kinds.

Probably one reason why broader economic co-operation has not been established with individual developing countries is Sweden's lack of a really suitable administrative apparatus for this purpose. The developing countries which have been interested in industrial co-operation have looked for an inter-governmental agreement, in which the Swedish government provides assurances of one kind or another that firms will honour their commitments. Sweden has not been able to meet this requirement, because the state cannot lay down the terms - prices included - to be followed by Swedish firms. Instead, the developing countries have been referred directly to the companies concerned. If closer economic co-operation with certain developing countries is judged desirable, a new form of international co-operation will have to be designed, in which the Government participates in negotiations and concludes agreements and, at the same time, binds Swedish firms to specific commitments.

The whole of this section has been concerned with possible co-operation between Sweden and various developing countries but it should also be made clear that consideration for our supplies of raw materials may prompt more active co-operation with countries nearer home. A Nordic[1] Commission on Resources and Raw Materials

(1) The word 'Nordic' refers generally to the countries Denmark, Finland, Iceland, Norway and Sweden.

is now in the process of defining fields in which Nordic co-operation can be widened. The raw materials assets of the Nordic countries are in many ways complementary, and there are reasons to consider the ways in which the Nordic countries may co-operate regarding raw materials, especially if one takes a gloomy view of the future.

Long-range Planning Perspectives

Uncertainty regarding future developments costs money, i.e. resources. This is most immediately expressed by the existence of a futures market on currency exchanges, where an exact price is put on uncertainties. In a previous chapter, we had cause to consider whether investments in the raw materials sector had been impeded by uncertainty regarding future government policy. As we have already seen, Sweden endeavours to reduce the element of uncertainty by maintaining relatively large wartime and peacetime contingency stocks, and these cost money. Even so, these stocks are only meant to protect the Swedish economy on a short-term basis; they afford no protection against more lasting changes in the conditions under which Swedish commerce and industry have to operate.

There are many ways of approaching the question as to how the state is to reduce uncertainty concerning the future. We have already mentioned tendencies towards bilateral agreements. Guarantees of raw material supplies of and markets for one's own products are one way, often at the price of higher raw material prices and diminished advantages in connection with the international division of labour. There are other methods too. For example, the state can guarantee prices, i.e. bear the brunt of uncertainty. Credit guarantees have the same effect.

Another, more far-reaching method of reducing future uncertainty would be too draw up alternative plans for commerce and industry based on different assumptions concerning the availability of raw materials and their prices. Thus proposals have been made in the United States to the effect that new mines should be prospected and invested in but then put in mothballs, i.e. left unexploited until prices go up or a physical shortage develops. Sweden could make similar preparations for the utilization of natural assets of various kinds, as a form of insurance against future shortages.

Planning, however, need not be limited to broadening the raw material base. It could equally well focus on opportunities of saving or replacing raw materials that were deemed liable to run short. This might be compared to stockpiles, insofar as saving would only be started if developments so required. The advantage of such a policy would lie in preparing the ground for a more permanent transition to raw materials that are abundantly available. Moreover, planning could be made to include measures aimed at more efficient recycling. The main gain, however, would be that changes to new raw material bases and new production processes would not have to be made at very short notice and without prior planning. The difference, compared with present-day research in universities and industry, would be that the main focus of attention was not on the raw materials and production methods that were cheapest at present, but on those that were certain to be available even in situations of crisis. Contingency planning of this kind could include, for example, steps to base the petrochemical industry on pulp-based raw materials, or to develop alternative alloying elements for different types of steel, or to encourage the more widespread use of ceramics, and to reduce energy and raw material inputs into agriculture.

Planning of this kind, needless to say, would also cost society a great deal. This would indirectly prejudice the aim of lower raw material prices, but on the other hand it would increase the scope to manoeuvre and safeguard employment.

Research

As we have seen in previous chapters, the corpus of scientific knowledge is the crucial factor determining which raw materials are extracted, which sources of raw materials are worth exploiting, or what amount of raw materials is consumed in the manufacture of a certain product.

Earlier we had occasion to state research requirements in relation to a particular topic of inquiry. Research is often described in relation to the raw material sector as a whole. It is only in this way that priorities can be determined between different kinds of research and a genuine research policy be framed. The German (BRD) programme of raw materials research is reproduced in table 19 as an example of this kind of policy.

The German(BRD) programme is a good reflection of the focus on natural sciences that is typical of most national research policies concerning raw materials. Increased efforts in the field of natural sciences can undoubtedly create greater liberty of action concerning the choice of raw materials and conservation measures. There is probably a historically conditioned tendency to refer to natural sciences in most such contexts because natural resources are regarded as something physical and biological. This is apt to result in a rather distorted view of problems which are more economic or organizational in character.

Tables 19. A German(BRD) programme of research into raw materials.

A. Aim: A wider raw material base	B. Aim: The economical use of raw materials	C. Aim: Increased recycling
1. Wider fields of prospecting (greater depth, unexplored regions)	1. Substitution of more abundant raw materials for resources that are scarce	1. Recycling and re-use of valuable material from used products.
2. More relaible prospecting methods	2. Utilization of all metals occurring in slag, etc.	2. Utilization of production waste
3. Prospecting of hitherto unexploited deposits	3. Reduction of specific material inputs	3. Product designs to facilitate recycling
4. Improvements in the exploitation of known deposits	4. Reduction of production losses	
5. Measures to improve the efficiency of mining and benefication techniques	5. Improvement of material utilization	
	6. More flexible use of raw materials, access to alternative processes	
	7. Improvements to the quality and service life of products - for example by reducing wear and corrosion	
	8. Multiple use of products	

We have already quoted examples of organizational problems connected with the collection of waste paper. The Nordic Waste Exchange is an organizational solution (being a means of sharing information concerning residual products) to a problem that is partly technological. Thus the problem consituted by the quantity of waste can be tackled either by attempting to reduce the volume of waste or by devising uses for waste. Presumably, therefore, greater concentration on the economic analysis or raw material flows and on problems of organizational theory, and economics connected with the product cycle, will open up new opportunities for saving raw materials or for using novel raw materials.

Summary

In this chapter, we have considered a number of instruments in a raw materials policy which have been used or contemplated in Sweden or other countries. The main impression made by this review has perhaps been a negative one; every measure we have mentioned has run counter to one or more of the societal goals which were initially discussed.

Should Sweden therefore completely relinquish the idea of a more active and more systematic policy on raw materials? We believe that this would be a premature conclusion. We have described different measures as frankly as possible. Since these instruments have been discussed in other countries and may become interesting in Sweden, it seems both natural and important that they should be discussed in this report.

The choice between national and international perspectives is important for a futures study. It should be clear by now that we have operated from two points of departure - Sweden's national interests and 'international solidarity'. We have pointed out a number of conflicts between these goals.

There are no instruments of raw materials policy which exclusively promote a policy 'in Sweden's interests'. Bilateral agreements concerning raw materials can be purely concerned with Swedish interests, but they can also improve the ability of the developing country concerned to plan on a long-range basis. The concept of longer planning perspectives can be employed in Sweden's own self-interest, but they could also be made part of an international policy of solidarity - if, for example, they facilitated the transfer of production to Third World countries.

The effects of policy instruments depend on the social frames within which they operate. This is equally true when considering possible means of resolving conflicts between policies of national interest and 'international solidarity'. An example of this could occur if output of a certain raw material was reduced in Sweden and handed over to countries with cheaper raw materials, while Swedish output concentrated on products higher up on the processing scale. Sweden would then concentrate on exporting high technology products (and possibly entire industrial complexes), and on exports of services, e.g. consultancy work on building up industries or administrations. Development in this direction could mean that the developing countries remained just as dependent on the industrialized countries as they are at present. But it could also assume such a guise as to assist the developing countries in their efforts to achieve independent development. Thus it is the political framework of technology transfer that will decide its impact on Sweden, and on the developing country or countries concerned.

We have also tried in previous chapters to show how uncertainty in the raw materials sector involves a variety of dimensions - economic, ecological, technological and political. Although these problems are so many-sided, and thus do not really lend themselves to short summaries, we still believe that some of our conclusions can be presented in condensed form.

It is both the material base and the social environment that decide whether something is a resource. Both these aspects have to be included in a presentation of the future resource situation.

The physical assets of the earth are so great that they can easily provide the volume of non-energy raw materials likely to be demanded for the period ending in the year 2000. (By then, the annual production of raw materials may be twice what it is today).

On the other hand, it is already possible to discern theoretical limits to a feasible production rate for renewable raw materials, especially fish and timber. Consequently, consumption of renewable resources will not be able to continue to increase very far into the next century. The crucial limit regarding the few metals and other substances which in the really long run may constitute the basis of civilization (e.g. iron, aluminium, magnesium, silicon) will be set by the energy required for processing, rather than the availability of the minerals themselves. On the other hand, certain metals will probably become exhausted some time during the next century, in the sense of becoming so expensive that a steep decline occurs in their extraction.

Thus the physical limits of resources are unlikely to constitute a primary problem for the remainder of this century. Instead, human living standards will hinge on the problem posed by the distribution of political power, and income.

This view is corroborated by the knowledge that social and institutional frameworks today are already causing people to starve despite the fact that the global production capacity in purely physical terms is more than capable of supporting the present-day population at reasonable levels.

The further one tries to look ahead, unless far-reaching changes of distribution are carried out, we will be exposing ourselves to steeply escalated risks in connection with the necessary extraction of raw materials. These risks comprise both ecological hazards and the danger of physical and mental strains. They also include the risk that our ability to shape a democratic society is circumscribed.

In our presentation, the graver consequences of this possible course of development have been envisaged as occurring (or potentially occurring) after the year 2000. This does not mean to say that they may be disregarded until then. In many cases, the process of social adjustment will be so long that a change will have to be initiated in what may be termed the near future.

The way in which tomorrow's resource perpectives will be developing will depend on the actions taken by individuals, organizations (e.g public opinion organisations and enterprises) and states. We hope that we have provided something of a background for a wider discussion of alternative courses of future action.

Appendix 1

The Secretariat for Futures Studies - Organization

Background

The discussion in Sweden during the later 1960s and early 1970s on the use of futures studies in governmental work lead to the setting up in 1971 of an ad hoc working group, under the chairmanship of Cabinet Minister Alva Myrdal. In 1972, this group published an official report entitled "To choose a future - background material for discussions and considerations on the role of futures studies in Sweden".

On the recommendation of the working group, a Secretariat for Futures Studies was set up in February 1973. It was attached to the Prime Minister's Office and was commissioned to continue the investigations on the role of futures studies, based on the responses to the first report.

Some changes were made in the organization of the Secretariat in accordance with a decision taken by Parliament in 1975. The Government now appoints the board of the Secretariat, consisting of members of parliament. In 1976, Kerstin Aner, Member of Parliament, was elected Chairman of the Board. All five parties in Parliament are represented on the Board. The Chief Secretary of the Secretariat, Professor Lars Ingelstam, is also on the Board.

Projects

In 1974, Parliament allocated funds for different projects for the first time. The following four projects were initiated in 1975 and were completed in 1977-78.

- Energy and Society
- Resources and Raw Materials
- Working Life in the Future
- Sweden in the World Society

The Secretariat for Futures Studies carries out the decisions of the Board by commissioning studies.

The final report from the project "Resource and Raw Materials" is the first project report to be published. Final reports from the "Energy and Society Group" and from the group on "Sweden in the World Society" were also published in 1978.

A new set of projects was started in 1978; these were on "Care in Society" and on "Sweden in a New International Economic Order".

Project Groups

A number of persons have been engaged for each of the projects to do the work. One person has been appointed as project leader to be responsible for the work of the group, for keeping time-schedules, planning the budget and to be generally responsible for the project. The project leader for "Resource and Raw Materials" has been Staffan Sohlman, Head of Department, Ministry for Foreign Affairs. Other members of the project group have been Tomas Bertelman, Ernst Hollander, Carl-Axel Olsson, Marianne Parmsund, and Uno Svedin. In the early stages, Arne Johansson, Lars Kristofferson, and Maj Rosander also participated.

The independent status of the project groups, and the fact that they alone are responsible for the project, has not prevented close and informal contacts with the Secretariat.

An inter-departmental reference group has been attached to each of the project groups. The reference group has no responsibility for the work of the project group, and serves only as a two-way channel for information between the group and the ministries concerned with the subject of the study. The reference group for the Resource and Raw Materials group has been chaired by the Under-Secretaries of the Department of Commerce, Bengt Dennis (1974-76) and Johan Nordenfalk (1976-77).

Appendix 2

Interim reports and other background material compiled in the course of the Resources and Raw Materials study

This final report is based on a fairly comprehensive volume of documentation, most of it comprising Swedish and international publications - reports, books and articles. Interested readers will find various source references in the publications listed below.

As can be seen from figure 1 in Chapter 1, a great deal of the work done by the project group was concerned with synthesizing the existing studies. Altogether, eleven interim reports have been published describing this work (of which only this final report has been translated into English). In addition, a special report by Professor P G Kihlstedt, entitled "Public raw material supplies during an energy shortage", has been published by the Royal Academy of Engineering Sciences. Professor Kihlstedt's report was compiled for the purposes of this futures study and for the futures study entitled "Energy and Society".

The interim reports can be roughly divided into four categories:

Discussions of physical limits: reports nos. 301, 304, 306, 311

Mainly technological report: nos. 303, 305, 309

Mainly social-scientific reports: nos. 302, 307, 308

Reports mainly centering on Sweden: no. 310

There were plans for the publication of further interim reports, and consultants were engaged, but for various reasons - lack of time being one of them - these plans were not brought to fruition. A number of the contributions received, together with the drafts compiled within the project group, are available to those interested from the Secretariat for Future Studies. It should perhaps be emphasized that the non-publication of these essays is not necessarily a reflection of their quality. On the contrary, many of them are highly valuable.

All of the interim reports have been obtainable by subscription at the truly generous price of US$6.00 Some of them, e.g. reports 301-303, are now sold out, but even these reports should be accessible to interested parties in Sweden, because a number of public libraries have subscribed to the series. Reports can be ordered at production cost from Liber distribution, Forlagsorder, S-162 89 Vallingby, Sweden.

Interim reports

Report no. 301
Resources and raw materials - some points of departure for a future study (172pp)
(The project group)
Contents include:
The background, purpose and strategy of the RoR study
The international raw materials market 1972-1975: problems and initiatives
Swedish raw material sectors
Swedish raw material output
Presentation of some international raw material studies
Appendix: Survey of terminology in the resource and raw material sector

Report no. 302
Some reflections on natural resources and economics (64pp)
(Dr. Carl-Axel Olsson)
Contents include:
The concept of natural resources
Scarcity of natural resources - an attempted articulation of the problem
The model of the economic cycle and the concept of material balances
Scarcity of materials and economic development
The classical models of scarcity
The scarcity models and reality
Explaining technological progress
Solow on natural resources
Value neutrality in economic theory
Towards a new economic discipline

Report no. 303
Remote sensing of natural resources (86 pp)
(Claes Pilo, D.Eng., the Swedish Space Corporation)
Contents include:
The development of remote sensing during the past ten years
Remote sensing technology. The present-day situation and bottlenecks concerning future development
Examples of the application of remote sensing in the resource sector. The present situation and the future
International aspects

Report no. 304
Renewable raw materials (176 pp)
(Main editor: Staffan Sohlman - Head of Department)
Contents include:
Population and resource pressure
Agriculture (by Erik Swedborg, the Federation of Swedish Farmers)
Forestry (by Reidar Persson, the Royal College of Forestry)
General structural problems (by Professor Erich Jacoby, University of Stockholm)

Report no. 305
Aluminium - a raw material chain (141 pp)
(Dr Sten Lindeberg, Dr Uno Svedin)
Contents include:
The various links in the chain (resources, reserves, extraction, processing: bauxite ore to aluminium oxide, production of primary aluminium, secondary production of aluminium)

End-uses of aluminium
Society's stock of aluminium
Resource consumption in processing
The trade in bauxite, aluminium oxide and aluminium
Power over the flows of aluminium industry
Job opportunities in the aluminium industry
Environmental questions
New technologies
Substitution
Sweden

Report no. 306
The sea and marine resources (162 pp)
(Main editor: Tomas Bertelman - Head of Section)
Contents include:
Marine ecology - marine pollution (by Asst. Prof. Bjorn Ganning)
The sea as a source of food (by Asst. Prof. Hans Ackefors)
Marine mineral resources (by Bo Hall, D.Eng.)
The law of the sea - its content and its relevance to the distribution of resources (by Dr Karin Hjertonsson)

Report no. 307(1)
Politics and raw materials in an international perspective (220 pp)
(Tomas Bertelman - Head of Section)
Contents include:
The interests of the consumer countries: means and ends
Countries and resources
Some "profiles" of supply policy: The United States, Germany (BRD), France, the EEC
Other important countries: Canada and Australia, the Soviet Union and China
Raw material questions in a Third World perspective

Report no. 308
Recovery of raw materials (78 pp)
(Main editor: Staffan Sohlman - Head of Department)
Contents include:
Waste and recycling in a community perspective
A short international survey
Waste and recovery in Sweden today and tomorrow. (By Goran Adolfsson and Egil Carlsson, K-konsult)
Problems and opportunities concerning paper recycling
Appendix: Material recycling and the service life of capital goods. (By Tomas Restad, Ph.D. (Econ.))

Report no. 309(2)
Substitution of resources (240 pp)
(Editor and main author: Uno Svedin, Ph.D.)
Contents include:
Substitution and history
Enforced substitution: examples of production of substitutes in Sweden during the 1939-1945 enclosure. (By Tommy Bengtsson, B.A., University of Lund)
An economic analysis of the substitution process. (By Professor Karl Jungenfelt, the Stockholm School of Economics)
A technological/scientific analysis of the substitution process. (By Professor Goran Grimvall, the Royal Institute of Technology, Stockholm)

(1) Summarized in Appendix 3a) of the present report
(2) Summarized in Appendix 3c) of the present report

A systematic analysis of the substitution process
Causes and effects of substitution
Substitution opportunities - some future perspectives

Report no. 310
Soil as a resource (110 pp)
(By Tomas Lindencrona, D.Eng.)
Contents include:
Conditions of land use
The functions of land
Conflicts concerning land use

Report no. 311(1)
Non-renewable raw materials (256 pp)
(By Ernst Hollander, B.Sc. (Econ.), with a contribution by Kerstin Anell, B.Sc. (Econ.))
Contents include:
Some basic features of world mineral production
Forecasts concerning the development of consumption
Reserves - a term with many meanings
Scarcity of non-energy minerals
Appendix: presentation of individual metals and industrial minerals

During an earlier phase, the project group also included Arne Johansson, Lars Kristofferson and Maj Rosander, as evidenced by certain of the sub-studies.

Reports published by other bodies but initiated by the RoR study:

P G Kihlstedt	Public supply of raw materials during an energy shortage. The Royal Academy of Engineering Sciences, IVA report no. 112, 1977
Goran Wall:	Exergy - a useful term in resource accounting (Chalmers Institute of Technology, Department of Theoretical Physics), 1977

(1) Summarized in Appendix 3b) of the present report

Appendix 3a

Politics and raw materials in an international perspective
Summary of a study by Tomas Bertelman (Report number 307)

It is not easy to summarize meaningfully a large volume of particulars about different raw materials, different countries and different kinds of raw material problems. Instead of a highly concentrated summary of the interim report, this presentation will be aimed at discussing the "major issues" in a political framework.

1. A historical perspective

A complex mutual relationship exists between raw materials and international politics, and this fact has always influenced raw material supplies in the industrialized countries. The industrial developments of the past 150 years have not only been borne up by a dynamic economic system, scientific revolutions and technical creativity, but have also been founded on a constant supply of large quantities of cheap raw materials. Whenever a raw material has become too scarce and expensive, new ways have been found of surmounting the problem.

Due to the economic and military hegemony of the European countries, the new raw materials required as a result of technical innovation in industry could be rapidly procured from the greater part of the world, and supplies kept pace with the growing demand that was generated by rising incomes.

One question worth asking - although it is practically unanswerable - concerns the extent to which the course taken by industrial development was determined by the European and American domination of Latin America, Africa and Asia. In about 1840, the American citizen Goodyear discovered how to vulcanize rubber. Only a decade later, the wheels of horse-drawn carts in the cities of America were shod with vulcanized rubber, and at the turn of the century the advent of the motor industry caused an astronomical rise in demand for rubber; between 1890 and 1910 there was a fantastic rubber boom in Brazil, which had practically a monopoly of rubber production. Prices soared, but suddenly in 1913 they fell to a quarter of their former level. The Europeans had secretly transferred rubber tree seeds to Ceylon and Malaysia, and those seeds had now developed into productive trees. Both in Brazil and Southeast Asia, armies of poor rural labourers were forced to work for a pittance at the arduous business of collecting latex from the rubber trees.

The ifs of history are usually regarded as barren ground in the social sciences, but we may nevertheless allow ourselves to conjecture what would have happened if the emergent industrial nations had not discovered the usefulness of the rubber tree, or if the Latin American and Southeast Asian countries had been unwilling to sell or had demanded too high a price for their rubber. Probably the horse-drawn carts would have kept rolling for a few more decades, and changed to dearer synthetic rubber (or some similar material) earlier than they actually did. Germany was already producing synthetic rubber during the First World War. The wheels of the Industrial Revolution might have turned a little more slowly, but they would hardly have ground to a halt.

From the mid-nineteenth century onwards, the productivity of European agriculture was improved with the aid of large quantities of fertilizer from outside, first in the form of guano and later in the form of saltpetre from Peru and Chile. Because of the predominance of British interests in those countries the two substances were exported cheaply and in large quantities. Exports of saltpetre grew uninterruptedly until 1914, eventually accounting for two-thirds of Chile's export revenues. Then the German laboratories managed to produce a competitive nitrate fertilizer which utilized atmospheric nitrogen. Together with other substances, the synthetic products ousted saltpetre from the market, and Chile's flourishing export trade collapsed in a welter of ghost towns and abandoned desert railways.

If the saltpetre and the droppings of the guano birds had not been discovered on the coasts of South America, how long would the impoverishment of European agriculture have continued? Needless to say, the availability of large quantities of cheap raw materials reduces the intensity with which substitutes are searched for.[(1)] Other raw materials could probably occasion similar reflections. The domination of the rest of the world by Europe during the nineteenth century enabled European industry to choose the cheapest of several alternative raw materials, and therefore probably had the effect of accelerating overall economic developments. But control of foreign raw materials was hardly of itself a necessary prerequisite for technical and economic development. The basic raw materials - iron and coal and various base metals - were available in Europe itself.

"Without the colonies and their preferential deliveries of raw materials, developments might perhaps have taken a slightly different course, but there is absolutely no doubt that the industrial revolution, once it had got underway, would still have been accomplished even without colonial raw materials and markets".[(2)]

It would be going to far for present purposes to elaborate on the significance of the caveat "once it had got underway" in the passage quoted above. Suffice it to say that the importance of imperialism for the development of the industrialized countries extends beyond their need for raw materials and markets. According to their different value premises, observers like Hobson, Rosa Luxemburg and Cecil Rhodes onwards have pointed to other aspects, ranging from the importance of the colonies in the initial accumulation of capital at the beginning of the "new age", through to their role during the nineteenth century in the resolution of the social crises that resulted from the pressure of population, combined with the maintenance of low wages and large profits, which were partly achieved by means of cheap imported food.

Returning to the context of raw materials, it is in fact arguable that without the colonial system (which for many purposes included the nominally independent countries of Latin America), research would have been speeded up and new substances would have come into use as raw materials sooner than they did. In Germany, for

(1) Substitution problems are discussed in Appendix 3c
(2) Marian Radetzki: A New International Economic Order? Stockholm 1976. (In Swedish)

example, a piston engine driven by solid fuel (in the form of carbon powder) had already been developed in the early years of this century, but low oil prices stifled it at birth.

But the colonial system was there, and it left its mark on the raw material supplies to the newly industrialized countries and on their mutual competition right down to our own time. The historical background is important because, even in the 1970s, it is still competition and relative prices (together with the institutionalized power factors behind them), and not just "physical properties", that determined what things are used as or developed into the "raw materials" demanded by the economy.

Some of the distinctive background to the present-day situtation can be summed up in the following historical thumb-nail sketch. When in 1884 Germany became the last major European country to begin acquiring colonies, Bismarck countenanced the enterprise reluctanly and solely on grounds or prestige. The advocates of German colonialism pleased the advantages of raw materials obtainable from the country's own colonies, but the government of the time did not regard this as a matter of absolute necessity.(1) Free trade, convertible currencies, and investments in the colonies of other countries, enabled Germany to obtain supplies on more or less the same terms.

Germany lost her colonies as a result of the First World War, after no more than thirty-five years. During the inter-war years, under the pressure of growing indebtedness, increasing trade barriers and currency barriers (with a franc zone and a sterling zone), Germany's demands on colonies was raised once more. This time, raw materials occupied the forefront. In an article published in Foreign Affairs in 1937, the German Minister for Economic Affairs summarized his country's demands in the following terms.

"Despite the League of Nations and its alleged assurances of peace, Japan has meanwhile decided to help herself and has acquired Manchuria; while Italy, by the conquest of Abyssinia, has expanded the territory which she requires for her life. As a result, Japan and Italy are no longer amongst the unsatisfied nations. They have left the Have-nots and have joined the ranks of the Haves, those nations which are satisfied. Germany remains the lone unsatisfied large Power. So long, then, as the problem of colonial raw materials is not solved for Germany, so long will she remain a source of unrest despite all her love of peace".

"It is either silly or cynical, in the face of such facts, for foreign commentators to declare that Germany can buy raw materials in the world market at will. No, Germany cannot do that because she does not possess the means of paying for them in foreign currencies; and she does not possess the means because foreign countries do not consume enough of her wares.

We know very well that even if we succeed in replacing with artifical native products a number of raw materials normally supplied by the world market we can do this only at high cost. So we ought to reject autarchy on principle, because it will necessarily lead to a lowering of the standard of life of the German people".

"A particularly ridiculous charge to which Germany has often to listen in connection with her colonial demands is that colonies in general and her former colonies in particular are valueless, and that it would not do Germany any good if her colonies were returned to her. This immediately prompts the retort: If the colonies are so bad, why do you keep them?"

(1) W.G. Henderson: Studies in German Colonial History, London 1962

"Today, when there no longer is free trade in the world, when Germany is crushed by foreign debts and harassed by the lack of raw materials and valuta, if her colonies were returned to her she would proceed to develop them with far greater intensity. A large part of the food supplies and raw materials which we now lack could be furnished by them".

"I therefore wish to name two conditions essential to the solution of Germany's raw material problem. First, Germany must produce her raw materials on territory under her own management. Second, this colonial territory must form part of her own monetary system".

"These, then, are Germany's two basic demands in the colonial field: that she have territories under German management and included in the German monetary system.

All the other questions involved - sovereignty, army, police, law, the churches, international collaboration - are open to discussion".(1)

Today, forty years on, these German demands appear as barbaric as they are old fashioned. Present-day conditions differ from those of the interwar years. Already in August 1941, in the Atlantic Charter issued after their meeting in the North Atlantic, Churchill and Roosevelt declared their intention that "they will endeavour to further the enjoyment by all States, great or small, victor or vanquished, of access, on equal terms, to the trade and to the raw materials of the world which are needed for their economic prosperity".(2)

Bearing in mind the German arguments during the interwar years, one is tempted to interpret the above section from the Atlantic Charter as a bid to prevent problems regarding raw materials from leading to a renewal of Germany's colonial demands. But the Americans too had a vested interest in gaining easier access to the raw materials of the European colonies. The post-war period has indeed seen a substantial rise in American imports of raw materials.

The international economic system developed after the war through the Bretton Woods and GATT agreements can be summed up in the following (simplified) terms:

- a multilateral liberalization of world trade
- free movements of capital, convertible currencies and fixed exchange rates
- the abolition of the "private hunting grounds" which the European colonies had constituted from the point of view of the United States (and Germany and Japan).(3)

Free access to raw materials, and a certain multilateralization of development assistance, were two steps on the road towards the abolition of the "private hunting grounds". Perhaps one can endorse the above-quoted report from the French Department of Industry in viewing the UN Decolonialization Declaration of 1960 as a continuation of the implementation of the principles enshrined in the Atlantic Charter. Among other things, the Declaration of Decolonialization calls for the abolition of the specific legal and economic ties between the colonies and their former suzerains and, accordingly, the admission of the new independent states to a more open international economy. At all events, the expansion of raw material

(1) Hjalmar Schacht: Germany's Colonial Demands, Foreign Affairs, January 1937 p. 228-234
(2) Foreign Relations of the United States 1941, Vol 1, Washington 1958, p. 368
(3) The term "private hunting grounds" (Chasses gardees) comes from a study published by the French Department of Industry in the series entitled "Etudes de la politique industrielle". "La Division Internationale du Travail", Vol 1, La Documentation Francaise, 1976, p. 127

investments in the developing countries by those industrialized countries which have not been major colonial powers (first the United States, and then West Germany and Japan) does not preclude such an interpretation.

This historical digression, incomplete as it is, serves to illustrate a number of tendencies in the raw material policies of the industrialized countries, and which can still be presumed valid and which will probably be of future importance. They can be briefly summarized as follows:

(1) The industrialized countries have an interest in obtaining supplies of raw materials as cheaply as possible, and they employ general instruments of foreign policy (alliances, intervention, foreign aid, etc.) to this end, insofar as they consider it feasible and appropriate to do so.

(2) The possibilities of finding new raw materials, new processes, new products and new applications in the industrialized countries are by no means exhausted. Developments in this direction are determined by profitability. This means that as long as raw materials can be obtained more cheaply by imports from abroad than by production at home, an increasing proportion of consumption will be imported, and less will be invested in research and development into alternative raw materials.

(3) An excessive difference between the different major industrialized countries concerning the cost of supplies of raw materials cannot be accepted. Even though the raw materials costs of individual countries can differ widely at present, free trade competition in finished products imposes limits on the permissible magnitude of such differences.

In addition to these - possibly self-evident - tendencies, a number of other factors have become increasingly plain. These include the following:

- The options open to the governments of the individual industrialized nations are rendered more complex by the fact that the transnational enterprises have strengthened their hold on many raw material sectors, and by the possibility that these enterprises have interests which substantially deviate from those of home governments and host governments alike.

- Environmental problems are acquiring both increasing importance and emphasis in agriculture, forestry, mineral extraction and energy conversion.

- General economic problems in industrialized countries (e.g. employment, regional development, balance of payments) can assume such proportions that they influence supply policy and prompt different measures from those which could be predicted in a system of free trade.

The extent to which these factors are allowed to influence the supply policy of individual countries depends on the general ideological tradition of each country regarding free trade versus controls, and also on their respective domestic raw material resources. There are great differences between the major industrialized countries as regards both these factors.

2. The 1970s

The great difference between the present-day situation and that which existed when Roosevelt and Churchill drew up the Atlantic Charter lies in the new position of the developing countries, the former colonies, and in their deliberate use of common action to obtain, first, better returns on their one-sided exports of raw materials and, second, to achieve a more diversified economic development.

The spectacular successes of the petroleum-exporting developing countries, the formation of a number of new organizations of producers or exporters, the Sixth Special Session of the UN General Assembly in 1974, the movement of the non-aligned states (with its increasingly radical demands on the industrialized countries, and its growing disenchantment with them), the raw materials boom between 1972 and 1974 - these and other events gave a powerful impression that the rules of the game were changing, and that the developing countries would actually be able to redistributed a portion of the world's riches by substantially raising the prices of their raw materials.

This possibility is easily exaggerated and easily underrated. The tremendous technological, industrial and economic superiority of the industrialized countries cannot be demolished at a stroke. On the contrary, this superiority makes them less vulnerable to the demands made by the developing countries in the raw materials sector. On the other hand, many of these demands were precipitately rejected by the industrialized countries in 1975 and 1976 as being unrealistic. The recession of those years reduced demand for raw materials and, with it, the prospects of exercising any producer power even on a short-term basis. This state of affairs may change. It should also be borne in mind that OPEC had existed for over ten years before it became an effective cartel. Co-ordination takes time, and little more than fifteen years have passed since the majority of developing countries were still colonies.

The demands which have been jointly pressed by the developing countries can appear contradictory. On the one hand, they demand to be allowed to export larger quantities of raw materials at better prices to the industrialized countries, in order to earn more foreign exchange and be able to import more technology, capital equipment and consumer goods from the same industrialized countries. On the other hand, they talk a great deal about self-reliance, i.e. emancipation from dependence on their own resources and focussing on their own inhabitants' basic needs in the way of food, clothing, housing, health and education. This would probably mean less commodity exports and a smaller but more balanced volume of trade generally.

This is how two critical African researchers have described the problem.

"What needs to be created is an integrated African economy oriented not to the needs of the economies of the West, but to the needs of Africa as defined politically by the African people. Anything short of that will prove incompatible with our aspirations for political and cultural autonomy". (1)

"The fault of the post-colonial policy was to continue the same kind of commercial farming to the detriment of the subsistence sector.

The Nigerian planners and policy-makers should therefore work consciously to change the content of our agriculture in such a way as to take care of the subsistence needs of the people. The use of more progressive techniques of production and marketing could be made available as much to the subsistence sector as to the cash crop sector, and greater emphasis should be placed on production for the domestic market rather than the international market.....

It will also be realized that as we gradually shift towards agro-industrialization, the volume of demand for local products will begin to expand, thus rendering export

(1) Chinweizu: The West and the Rest of Us, New York, Vintage Books 1975, pp 286-287

production ill-advised. The soap and rubber industries, to mention only two, can buy up all the palm oil and rubber produced in the country".[1]

But the developing countries are not uniform entities. Indeed, they greatly surpass the industrialized countries in terms of injustices, oppression and economic inequality. A change to self-reliance, the termination of the externally oriented economy based on raw materials, which has so often meant subservience and under-development instead of self-propagating development, would require profound transformations of the majority of Third World societies. Those transformations would not favour the local elite or upper class represented by many of the Third World governments - city dwellers living on exports of raw materials from the countryside. As long as the present social structure survives in the majority of developing countries, the industrialized countries need hardly fear any decline in their exports to the Third World as a result of resolutions about economic development characterized by self-reliance and reduced trade.

Despite the community of interest between enterprises in the industrialized countries and the governing strata of the developing countries, there are of course conflicting interests involved here as well. The governments of the developing countries want to earn more on their externally oriented economy and to be better paid for their raw materials, and they want a larger proportion of value added based on their own raw materials to accrue to their own countries. These are the main topics of discussion, both in the intermittently tough negotiations between raw material enterprises and individual developing country governments, as well as in the more general proceedings of UNCTAD and other multilateral fora.

How, then, do the major industrialized countries react to the situation which has arisen during the 1970's? Most of them have begun studying their raw material supplies, and those studies have furnished most of the material for this report. Firstly, it is clear the opportunitues for self-sufficiency vary a great deal between the four largest OECD countries. Whereas the United States has an abundance of most minerals, timber and foodstuffs[2], Japan is ostensibly lacking in all three. Germany (BRD) and France occupy an intermediate position, where minerals are concerned, and even this means heavy dependence on other countries. The last major raw materials study in the United States showed that the country's own resources as regards more than two thirds of the 67 minerals studied were large enough to meet at least 75 per cent of national demand for the rest of this century. In many cases, reserves were several times greater than total demand. But prices will have to be raised above their present level in order for all these resources to become economically viable, and there are a number of other factors which will help to decide whether they are actually exploited or not.

Although "wealth per acre" differs a very great deal from one country to another, differences in national policies of supply are not determined by this factor alone, but by factors of an economic and general ideological nature. There is nothing odd about this. "Wealth per acre" defines the outer limits of the self-sufficiency attainable by a specific country, but it is the quest for cheap raw materials, and the other considerations mentioned previously, which, within these limits, decide a country's degree of self-sufficiency and the general shape of its raw materials policy.

A number of common features are clearly apparent. All industrialized countries assume, more or less explicitly, that consumption of raw materials will increase

(1) Marin Igbozurike: Problem-Generating Structures in Nigeria's Rural Development, the Scandinavian Institute for African Studies at the University of Uppsala, Uppsala 1976, p. 126

(2) As explained previously, fuels are not specifically dealt with in this report.

throughout the remainder of this century. For most minerals, the increase is expected to be less than that of the economy as a whole (with the exception of aluminium and a number of ferro-alloying elements). But even a moderate growth rate of 2.5-3 per cent per annum will mean a doubling of consumption in twenty-five years. France excepted, the major industrialized countries also anticipate that a growing share of consumption will be imported and that a growing share of their imports will come from the developing countries. France aims to increase her domestic output of minerals from 14 to 20 per cent of total consumption during the ten years between 1975 and 1985, and yet she still expects to import increasing amounts from countries outside Europe, primarily in Latin America.

In all countries, materials research is being intensified and efforts are being made to find substitutes for raw materials which cause problems. France and the United States, for example, have experimental plants for the production of aluminium from alternatives to bauxite. France is also making deliberate efforts to replace copper with other materials, though this is mainly because copper is a heavy burden on the French balance of payments, not because it is a particularly "risky" metal.

The United States, France and Japan have all resolved to establish what are termed economic stockpiles of raw materials. Economic stockpiles, as distinct from strategic stockpiles, are intended for use in peacetime crises due to, for example, producer cartels, international conflicts in other parts of the world, strikes, embargos and so on, and they are not designed to last for such long periods as strategic stocks. Strategic stocks in the United States are already very large, corresponding to half the world's annual output of certain materials. Stockpiling targets were recently raised still further. At present the strategic stocks of the United States are valued at some 7,500m. dollars. American economic stocks will probably come to exceed by far 1,000m. dollars. The target for French economic stocks has been set at 1,200m. francs. Although the size of the Japanese economic stocks is not known, we may safely assume that the combined stocks of the three countries will amount to more than 3,000m. dollars. This can be compared with the international buffer stocks proposed by the developing countries within the framework of the UNCTAD integrated raw materials programme, which are estimated to cost about 6,000m. dollars, and to require an initial contribution totalling 1,000m. or 1,500m. dollars; so far, this proposal has been turned down by the major industrialized nations. The strategic and economic stockpiling activities of the industrialized countries themselves can thus amount to a very important power factor, both as regards raw materials and concerning other aspects of international relations.

If stockpiling is mainly concerned with unforeseen events and with countering measures that may possibly be taken by the developing countries, the opportunites of export control, which the large industrial nations have acquired for themselves, are principally aimed at other industrial nations. One purpose of export controls is to prevent raw materials needed for the national industry from being drawn out of the country by higher prices offered by foreign consumers. In situations of crisis, the aim to prevent the occurrence of a physical shortage. In more normal situations, the aim is simply to restrain the level of domestic prices. The United States has imposed export controls, for example, on copper scrap, nickel and - perhaps the most widely known instance - soya, which has been a serious blow to Japan. The French are discussing possible means of putting an end to their heavy exports of copper scrap to other EEC countries.

The anticipated need for increased imports of raw materials by the industrialized countries from the developing countries has prompted many different measures aimed at stimulating output of minerals in the Third World. It is to be expected that foreign intelligence coverage will be stepped up through the appointment of special "mineral attaches" in interesting developing countries. Through their access to remote sensing data, and through their technological lead in the mining sector, etc., the big transnational corporations have a head start where prospecting in the developing countries is concerned. The governments of the major industrialized countries are endeavouring, by means of favourable tax regulations, credits, investment insurance and investment agreements, to induce companies to take the risks entailed by major extraction projects in the developing countries. The efforts made in this direction by all four of the major industrialized countries include international development assistance as one of several instruments, either in the form of geological expertise or else as long-term credits for mining facilities, general infrastructure aimed at facilitating the extraction of raw materials, or even a certain amount of "cultural assistance".

It is above all the raw material investments of the American transnational corporations that have stagnated or diminished in the developing countries during the 1960s and 1970s. It is not possible to arrive at a clear picture of the future plans of the North American raw material enterprises in the developing countries, but France, Germany (BRD) and Japan are all planning for a further expansion. On the specifically bilateral plane, this development appears to be proceeding more smoothly than one might be led to be believed by the emphasis in multilateral discussions on nationalization, hostility to investment in the developing countries, and the risk of insufficient expansion of capacity. Germany (BRD), at least, has reported that in fact it is quite easy to reach agreement with selected developing countries, and that it is possible for German (BRD) firms to gain a foothold in cases where the transnational US companies have left a gap for them. A study by the Japanese Department of International Trade and Industry, estimates that Japan's investments in the raw material sectors (agriculture, fisheries, forestry, mining, timber and pulp) will have multiplied nine times over in real terms between 1975 and 1985, rising from 4,300 to 34,400 million dollars.(1)

It is with regard to foreign investments that the industrialized nations have displayed the greatest interest in international co-operation to guard against future uncertainty in the raw materials sector. The United States in particular has suggested an international resource bank, on a multilateral basis, in order to help reduce the uncertainty of enterprises in the developing countries.

Extensive inter-governmental direct agreements concerning deliveries of raw materials (possibly geared to industrial co-operation) are relatively uncommon, and are mainly concerned with strategically important commodities (such as oil, coal, wheat and uranium). Long-term agreements for periods of between five and twenty years, however, are becoming increasingly common between individual companies. It is widely expected that inter-governmental involvements in the raw materials trade will increase, not only through UN negotiations but also bilaterally and possibly regionally.

Another point of similarity between the major industrialized countries is that they have succeeded for over a decade in resisting the most important demands put forward by the developing countries in politically coloured UNCTAD meetings and subsequent "dialogues". In this respect the United States, Germany (BRD) and Japan constitute a hard core in the industrialized camp, heavily underlining the importance of the mechanisms of the market being allowed to govern international exchange. They reject measures aimed at influencing supply or prices in the longer term. Yet all three of them have participated in a couple of international raw material agreements to demands for market controls, and has long advocated what is termed "organization of markets". Where Japan is concerned, support in principle of free

(1) Japan's Industrial Structure 1975 - A long Range Vision. Japan External Trade Organisation 1975, p. 29. Figures refer to fixed dollars at the level of prices in 1970.

markets is combined with a strong interest in bilateral long-term agreements and a willingness to accept slightly higher prices from time to time in return for greater security of supply.

At the multilateral level, the major industrialized countries have been equally successful during the UN Conference on the Law of the Sea in countering proposals to place the exploitation of the international seabed under UN control. The United States, Japan, Germany (BRD) and France are the main countries dominating the technology involved in the extraction of deep sea nodules, the contents of which (nickel, copper, iron and manganese) are all at present imported from the developing countries, though to differing extents.

In addition to the three centres of the OECD area (the USA, the EEC and Japan) and the developing countries, there are four countries which between them may come to have a great influence on the future raw material situation: Canada, Australia, the Soviet Union and China.

Canada and Australia are two large, sparsely populated, industrialized countries, both very rich in minerals and both economically oriented towards an industrial centre nearby (the United States and Japan respectively), which also account for the greater part of their exports. These exports are dominated by unprocessed raw materials, and two-thirds of the mining output of each country are in foreign hands. A great deal of the investments made during the 1960s and 1970s have gone to Canada and Australia instead of unruly developing countries. Investments in Canada and Australia have been judged a far safer proposition, because they are industrialized countries and share the same values as the rest of the OECD area. Nonetheless, both countries have shown tendencies during the 1970s towards a desire to exploit their own resources more in their own interests and less in accordance with those of foreign investors.

Despite an awareness of these contrary interests, there is really nothing to suggest that Canada and Australia will depart from their position as the mineral store of the western world. Instead, they envisage a steep rise in their mineral exports for the remainder of this century. Canada, however, is clearly endeavouring to establish domestic processing and national control. If for any reason a change were to occur in the export policies of these two countries, supplies in the rest of the OECD area would be radically transformed.

The Soviet Union and China are the other two countries which come outside the picture of exporting developing countries and importing industrialized countries. Both have very large potential mineral resources, and few observers doubt that they could become virtually self-sufficient in the future, if they wanted to. At present, both countries export oil as well as certain other mineral raw materials, most of the Soviet Union's exports going to eastern Europe. They also import a certain amount of minerals, China's main imports being base metals.

The rate at which these countries exploit their own assets is determined partly by political choice concerning a suitable degree of self-sufficiency and international economic integration respectively, and also by the availability of foreign exchange with which to acquire technology and capital equipment for the mining sector. Continuing exports of oil, and possibly other raw materials, to the OECD countries can be expected for the purpose of financing purchases of this kind.

But the Soviet Union and China occupy something of an intermediate position between industrialized and developing countries in the raw material sector. At the same time as they sell certain raw materials to the industrialized countries, they import others, mainly from the developing countries. Therefore one cannot entirely discount the possibility of their acquiring "import interests" vis a vis the developing countries and, at the same time, contributing (if only to a small extent) towards the competition for raw materials from the developing countries - as evidently already exists between the United States, the EEC countries and Japan. If China and the Soviet Union become integrated with the international economy, they will probably increasingly tend to demand the same commodities as other industrialized countries from the developing countries. Of course, this type of development is not predestined, and the only point in referring to such a possibility is to show that "independence" with regard to raw material supplies can very well exist in geological and technological terms without necessarily becoming an economic reality.

This concludes our simplified and highly condensed review of the geography of raw materials policy with particular emphasis on the non-energy minerals sector. In the context of domination and dependence, raw materials present many difficult problems which cannot be treated within the framework of a report of the present kind. But there are two questions which we will now consider briefly, namely the nature of the dependence of the industrialized countries, and the consequence of wastage and conservation of resources in a politico-economic context.

3. The nature of the dependence of the industrialized countries

What is the significance of Western Europe, the United States and Japan importing three-quarters of their consumption of more than a dozen important minerals, and of perhaps half their consumption of an even greater number? First of all, there are some distinctions that have to be drawn. Are we concerned with short-term or long-term dependence, and are we referring to the industrialized countries as a group or region by region? And finally, what assumptions do we make concerning Canada and Australia?

In the very long run (50-100 years), the industrialized countries may perhaps develop a technology which in principle is independent of supplies or raw materials from the developing countries, regardless of whether this is accomplished by new methods of extracting very small mineral contents or by the development of completely new materials (ceramics with special properties, etc.). Research has begun in both directions. The greatest optimists are already talking about the Age of Substitutability, when it will be possible to construct societies not very different from those existing today by exploiting the average contents of the five commonest minerals in the crust.(1) But the uncertainty attaching to the outer limits in the form of energy, climate, environment and large-scale social organization is so great that it cannot be adequately described, even in a special report. What we ought to emphasize instead is that very great changes will probably have to occur in the international political and economic structure and within individual countries before the industrialized countries seriously endeavour to make themselves independent, and also before the outer limits come to be regarded as common to all countries.

A generalised answer will also have to be given to the question of the short-term dependence of the industrialized countries regarding raw materials. In the first place, the imports of raw materials by the industrialized countries from the developing countries (their raw material dependence) are a historical result of their dominant position, not of their weakness. Thanks to their superiority in terms of military power, technology and organizational ability, they have been able to obtain, at low cost, tropical agricultural produce and to exploit mineral deposits in the developing countries after exhausting deposits of their own with high mineral content. This process has had profound social effects on the developing

(1) H.E. Goeller/A.Weinberg: The Age of Substitutability, in A Strategy for Resources, A Science Policy Foundation Symposium, Eindhoven, September 1975.

countries. Among other things, in the majority of cases a "modern" elite has come to power with a strong interest in perpetuating the extrovert, export-oriented economy of the developing countries. The dominant position of the industrialized countries continues, even though the former colonies have gained their independence, and for as long as this position can be maintained, imports of raw materials from the developing countries will probably continue to be both possible and profitable. It also seems clear that the industrialized countries have actively contributed during the twentieth century towards the perpetuation of this mutual dependence - the reproduction of the system. There have been several interventions aimed at preventing power getting into the hands of those advocating a more introvert, nationalist economy in the developing countries.

If then, most of the inustrialized countries are counting on importing a growing share of their consumption of raw materials during the next few decades, their raw material dependence will increase but, at the same time, they must be confident that their dominant position will remain, i.e. that the dependence of the developing countries will also continue.

Does the increasingly current term "interdependence" convey a fair picture of the conditions thus described? Assymetrical interdependence would seem to be a more apt expression. The dependence of the developing countries is so multidimensional that it cannot be compared with the need of the industrialized countries for imported raw materials. "Interdependence" would perhaps be a more accurate term if it referred solely to conditions in the trade in raw materials and emphasized that the industrialized countries needed raw materials, but the governments of the developing countries were just as interested in the export revenues.

The second question concerning dependence focuses on the short run magnitude of these revenues. Are the industrialized countries susceptible to abrupt sharp curtailments of deliveries by the developing countries, aimed at raising prices? Attempts of this kind would no doubt be capable of disrupting the economy of the OECD countries, breaking business cycles and fuelling inflation. But the industrialized nations, having placed themselves in this vulnerable position, probably still regard it as acceptable, for at least two reasons. The first reason is described in the following terms by the Iranian economist Jahangir Amuzegar, one of the delegates at the CIEC in Paris.

"The "new cold war" is neither essentially ideological nor truly systems-oriented. It is, for the most part, a simple bread-and-butter issue in the fundamental sense of the word. The Third World is not opposed to economic growth either for itself or for the industrial world. As a matter of plain common sense, the only way the poorer countries can ever catch up with the richer nations is to grow a lot faster, continue to consume a lot less, and invest a lot more. They can thus hardly be expected to oppose their own economic growth. And since their material progress is closely intertwined with Western prosperity, they would be fools to wish to kill the goose that lays the golden eggs".(1)

Seen in these terms, the short-term vulnerability of the industrialized nations could perhaps be described as a source of strength. It is in the interests of the Third World governments to avoid measures which would boomerang.

The second reason why the short-term vulnerability entailed by heavy dependence on imports appears acceptable to the industrialized countries is that the develop-

(1) Jahangir Amuzegar: The North-South Dialogue: From Conflict to Compromise. Foreign Affairs, April 1976. p.557-558. The expression "bread and butter issue" dates from the period when trade union militancy was concerned with higher wages and did not refer to questions of power and influence.

ing countries have relatively little chance of converting their potential power into actual power. Differences and conflicts of interests between the developing countries themselves, an economic situation which is often strained and lacks the extra resources for "campaign funds" and, not least, the ability of the industrialized countries to divide and rule by means of discriminatory agreements, development assistance and threats of intervention - all of these things help to weaken the power of the resource weapon even in the short run. In addition, more and more industrialized countries have started to build up economic stockpiles for the very purpose of reducing this type of short-term vulnerability.

The consumption by the industrialized countries of a large quantity of raw materials from the developing countries can be referred to as dependence, but there is no straightforward connection between imports as a share of their consumption and their vulnerability. In fact, it is not easy to see how, within the existing framework, the developing countries could employ their resources to inflict any serious damage on the industrialized countries - at least, not as long as the latter have both the resources and the attested preparedness to take countermeasures.

Suppose that the exporting developing countries were really able (with tacit consent of Canada and Australia) to up the prices of a dozen or so of their minerals to twice or three times their former level. What effect would this have? Perhaps the growth of the large OECD countries would decline somewhat and their inflation rise somewhat; perhaps the balance of payments in a number of countries would deteriorate further. New mineral production would become profitable in several industrial countries. There would be a transfer of real resources to a limited number of developing countries, but the fundamental economic structure of the latter, dependent as it is on exports of raw materials, would be unaltered. The probability is that higher raw material prices will neither break the industrial nations nor make the developing ones.

Rising raw material prices can, on the other hand, have various consequences in various parts of the industrialized bloc. Higher raw material prices mean increased extraction of those deposits that were not viable previously. This will above all favour the United States at the expense of the EEC and Japan. It has been pointed out in various quarters that a rise in the costs of imported raw materials and energy would reduce the competitive strength of the European and Japanese economies in relation to the United States. This would probably go unlamented in the United States, whose trade in finished products (armaments excluded) is showing a mounting deficit, and whose consumption of finished goods includes a rising proportion of imports. There has been speculation concerning the extent to which knowledge of this fact has influenced the actions of the United States in matters relating to raw materials. There are no clear indications that raw material prices have been deliberately manipulated by the United States to these ends. On the other hand, it is possible that the greatly inferior supply position of the European countries will lead them in the future to regard possible global settlements as insufficient partly because such settlements may be dominated by the interests of the United States. Instead they will prefer to deepen the bilateral or regional co-operation which has been proceeding under the agreements concluded, first, with the African countries and then, in 1975, with the ACP countries (i.e. the former European colonies in Africa, the Caribbean and the Pacific). Japan, together with East and Southeast Asia, could become involved in similar developments.

The above discussion does not provide a complete answer to questions about the nature of the raw material dependence of the industrialized countries. It should serve to show, however, that before pronouncing on the relationship between imports of raw materials by the industrialized countries and their "vulnerability" today and tomorrow, one has to analyse, among other things, the world power structure.

4. The effects of wastefulness and conservation

The pattern of conflict, domination and dependence within and between countries also determines the effects of what, in the debate on resources, are termed to waste or to conserve. Space will not permit us to analyse the debate in full, but there are two features of it which we may consider. Arguments in favour of conservation (here roughly defined as a deliberate reduction of the consumption of material resources) can perhaps be divided into two groups, namely egoistic motives and, secondly, motives of solidarity or justice (motives of "enlightened self-interest").

Environmental considerations probably feature largest in the first group. The argument that we must economize on raw materials, because otherwise they will run out, is a common one, though it is none too credible on closer inspection. From an environmental viewpoint, however, it is not only the extraction and processing of raw materials that cause serious pollution; the use and destruction of finished products also do so. Processes can from closed systems, materials recycled and so on, but it still seems reasonable to suppose that the more raw materials we put into the economic system in the long run, the greater will be the pressure exerted on an environment that is already under severe strain.

Another argument of a more long-term economic nature has been propounded in at least the French debate. This argument says that, in countries whose trade balance has a structural deficit (and France is one such country), imports of raw materials will become increasingly large, and competition between the three big industrial blocs to find markets for a growing surplus of essentially similar products from the manufacturing sector will be intensified. The Japanese economic forecasts seem to support this argument. In order to finance their imports of raw materials, the Japanese are counting on a trade surplus vis a vis Western Europe of no less than 17,000m. dollars by 1985. Even as it is, Japanese exports are regarded as a problem by the EEC. To countries like France, therefore, the presumed shortage of foreign markets for their exports would justify a more sparing use of raw materials.(1)

In Sweden, however, it is mostly considerations of justice that have dominated the debate on conservation of resources. More and more people are rightly coming to regard the uneven distribution of the earth's consumption of resources as increasingly unacceptable. Even among those who do not subscribe to equality as an absolute value, there are many who see a new source of future wars, and possibly of global destruction, in the widening gap between rich and poor countries and in the growing distress. In other words, enlightened self-interest argues for the solution of the new social question (which now refers to global relations instead of the working classes of late nineteenth century Europe, which originally prompted the expression).

There are two quite incompatible schools of thought on justice and the resource consumption of the affluent countries. One of these is represented by the speech made by Dr Kissinger at the 1976 session of UNCTAD: "Increasing American demand for products of other countries will make a major contribution to recovery around the world". Nobody at the Conference seemed disposed to deny this. "The richer we get, the better it will be for the poor countries", seemed to be a view that was shared by most of the other industrialized countries.(2)

(1) Jean Matouk: Pour une Nouvelle Croissance Implosive. Alfred Sauvy: Influencing Patterns of Production and Consumption.
(2) The two quotes from Kissingers speech given in this paragraph are given in John Tanner's article 'UNCTAD V' Resources Policy, September 1976, p. 211

The contrary opinion has many adherents in the Swedish debate: we must go without so that they will have access to resources. Georg Borgstrom, the Swedish scientist now working in the United States, has expressed this in many ways. Here is one example:

"The world economy cannot recover and man's destiny cannot be guided in a better direction until we come to terms with the one thing that is needed, namely equilibrium with the rest of the world. Land, food, forest and energy resources must be reserved for sane management, with all members of the world community being counted and being acknowledged as entitled to have their minimum wants provided for, at the same time as the rising tide of population is stemmed."(1)

Thus the questions is: are we to save in order to help the poor or are we to consume still more so as to help the poor? The answer depends on how one saves and on the time horizon and the social approach one adopts. It is only when these perspectives are made clear that words like "sane", "equilibrium" and "all members of the world community" can acquire concrete meaning.

If we refrain from part of our real income and transfer it to the developing countries, this will mean a reduction of our own consumption and, therefore, a kind of saving. The offical development assistance, which is financed out of taxation, performs this kind of function, and it also has the advantage that it can be directed towards countries and sectors which make it meaningful for developmental purposes.

There are other ways of abstaining or saving, besides paying tax for international assistance. For instance, the individual may decide for himself to consume smaller quantities of meat and other food, less energy and fewer products containing imported raw materials. In the short run, this will mean a decline in the export revenues of the developing countries and in their ability to import products from the industrialized countries. Consequently, the governments of the developing countries, wishing as they do to increase their export earnings, are disturbed by this type of saving. They do not want the industrialized countries to become self-sufficient in those products exported by the Third World, they do not want to see synthetic products which, for example, can take the place of rubber and hard fibres, and they want the industrialized countries to do away with taxes restraining consumption of their products (coffee tariffs, for example). Whereas the industrialized countries are pinning their hopes on the mineral resources of the sea bed,the developing countries are concerned by the thought of this rival source of raw materials, and so on. This difference of perspective between industrialized and developing countries reflects the state of dependence of the latter. If the industrialized countries begin to manage without their resources, this may cause the developing countries to become increasingly "marginalized" and to lose what importance they now have. Anybody wishing to help the developing countries by reducing his own consumption must adopt some sort of a standpoint with regard to this dependence and to the way in which "all members of the world community" form part of a structure of groups, classes and nations.

If, for example, all the industrialized countries stopped using soap made from palm oil from the developing countries, what would happen to the producing countries? In the short run, all of their inhabitants, from landowners to poor farmworkers, would probably be worse off. The distribution of the burden would be decided by the political situation in each country. The poor would be worst hit, insofar as their ability to cut down on their consumption is already slight.

(1) Georg Borgstrom: The Banquet, Stockholm 1975, p.8

What would happen in the longer run? It is impossible to say. If the produce of the plantations could not be sold in the world market, the labourers might possibly assume power and start growing things for their own immediate needs and those of their country. If the industrialized countries had no demand for the produce of the plantations, forests and mines, perhaps their governments would have no cause to intervene in the developing countries in defence of the owners of plantations, forests and mines, and in opposition to the attempts of the poor to use the land for their own needs? This discussion is partially reminiscent of that which customarily ensues in Sweden concerning suggested tourist boycotts of dictatorships, boycotts of goods from countries like Rhodesia, South Africa, Chile and so on. But the question in these cases is purely one of a boycott; nobody thinks of it as a form of saving.

A development of this kind seems unlikely to materialize as a result of saving in the industrialized countries. But what about the reverse? Can saving possibly have the effect of making the industrialized countries less disposed to retaliate in future if they are confronted with political upheavals in the developing countries and with new regimes which for political reasons, do not wish to perpetuate their economic dependents i.e. on exports? Viewed in this perspective, perhaps saving can be a long-term process of preparation in the affluent countries for possible changes in the international economic system. But there is nothing in history so far to suggest that the actual change will be initiated by the industrialized countries themselves. Privileged groups and countries have never relinquished their privileges without being forced to do so.

This discussion does not profess to be exhaustive, but it is meant to show that one has to state some viewpoint concerning what makes the world tick in order to be able to indicate the effect of measures one proposes. Otherwise, what is the difference between the Project Independence proposed by President Nixon as a countermeasure against the power of the OPEC countries, and saving prompted by considerations of solidarity, be it saving of foodstuffs, energy or minerals? The effect may be the same, even if the underlying purposes are completely different.

The world is full of economic contradictions and conflicts which make it hard to judge the effects of actions taken. When Joseph advised Pharaoh to store food for the lean years, nobody could have thought that his advice was anything but sensible. But not even this precautionary measure remained a purely Egyptian concern.[1] On the contrary, people came from the land of Canaan, where there was famine, and "fell down" before Joseph, who was able by virtue of his "agripower" to induce Jacob to surrender his youngest son. For similar reasons, the well known think tank of the Battelle Memorial Institute wrote as follows in a report to the US Government in 1973, before the major industrial nations decided to join the USA in building up economic stocks of raw materials designed to last, if not for seven years, then at least for one.

"When a foreign nation signals even the slightest intention of entering into stockpiling as a national policy, it raises questions by others about its purposes, intentions, and management policies. Questions as to whether stockpile management decisions and tactics will be stabilizing or unstabilizing, or a threat or a boon to one's own national interest are of great concern. In the absence of clarification, the prudent reaction is to assume the worst and take the act at first as an aggression and a threat".[2]

(1) Genesis XLI-XLII
(2) Battelle Memorial Institute, Report 1973, cit. "US Dependence on Imports of Five Critical Minerals: Implications and Policy Alternatives", p. 75 General Accounting Office, 29th January 1976.

5. Sweden's role

This report is concerned with the main global relationships between raw materials and politics, and hence the small amount of attention that has been devoted to Sweden. By way of conclusion, however, it may be interesting to point out one or two distinctive features characterizing Sweden's position.

It is hard to see why Sweden's supplies of raw materials should be subject to different conditions from those applying to the industrialized countries generally. If the same analytical model were to be applied to Sweden as to the other industrialized countries, it would probably show that free trade and multilateral commercial contacts with the developing countries are thought to be in our interests, and that our own large exports of raw materials can make certain forms of producer co-operation attractive. Owing to her heavy exports of forest products, special steels and ores, Sweden is classified within the OECD as a net exporter of raw materials. But our imports of alloying elements, tropical foodstuffs and other agricultural products are obtained on the same conditions as those of other industrilaized countries. This is virtually self-evident, and it merely reflects the simple fact that Sweden is a part of the industrial bloc in the world.

The industrial bloc, however, is not a homogeneious unit, and the same goes for its relations with the developing countries. For example, Sweden has gone further than most other industrialized nations towards meeting the demands of the developing countries regarding the global issues of development assistance and trade, and Sweden has also been more explicit than most of the affluent countries in declaring her support for the demand of the Third World for a New International Economic Order. Many of these demands nevertheless remain unfilled, even where Sweden is concerned, and Sweden, like other industrialized countries, has benefitted from the rules which have been laid down for the global economy, i.e. the very rules which the governments of the developing countries wish to modify. The fact also remains that the major industrial countries, in seeking to maintain these rules, have resorted to measures which Sweden rejects on grounds of principle, but the Swedish economy has still been able to derive benefit as a result of them. On the other hand there is not much that a small country can do on its own - even if it would like to - in order to change the rules of the game, i.e. the rules of the international economy.

With this mind, what standpoint should people in Sweden, wishing to contribute towards greater international justice, adopt concerning our supplies of raw materials in an inequitable world? Our earlier discussion showed that "saving" or "conservation" is not an unequivocal or self-evident alternative, least of all in the short run.

Nor does it seem to be a self-evident recourse to work generally for higher raw material prices, so as to "help the developing countries". As development assistance, resource transfers of this kind would be a blunt instrument. Distribution between and within the developing countries would be very uneven, and an unnecessarily large portion of the resources would probably be spent on purposes with which we do not sympathize. Several industrialized countries would also be gratuitously favoured. Moreover, the arguments against higher raw materials prices, on the grounds that they would be liable to cause overproduction and the development of substitutes, are not just trumped up objections; they reflect economic realities.

Sweden can contribute to international raw material agreements which stabilize prices in raw material markets, or to arrangements aimed at stabilizing the revenues derived by developing countries from their exports of raw materials. Mechanisms of this kind would improve the ability of the developing countries to conduct long-term economic planning of their own, and would therefore be highly valuable. In recent years, Sweden has in fact initiated a global programme for stabilizing the export revenues of the developing countries. A programme of this kind would be useful, but it would only be a very small step towards the establishment of a more equitable world. Insofar as it meant a transfer of real resources to the developing countries, there would always be the risk of a corresponding deduction being made from the other assistance provided by the industrialized countries. Sweden's contribution to the transfer of resources via trade in raw materials should not be made at the expense of direct Swedish assistance to the developing countries. Increased direct Swedish assistance, focussing on countries deliberately following a development policy aimed at social and economic justice and at the needs of the majority of the people, is still the best contribution that Sweden can make towards reducing inequalities in the world. But this is by no means a new discovery in the internal Swedish debate.

Even if the origin of the economic underdevelopment of the developing countries is to be sought in their enforced specialization in raw material exports, it may be that (as we have already seen) this underdevelopment cannot be broken by our purchasing smaller quantities of raw materials or paying higher prices. Underdevelopment can only be cured by the developing countries themselves. The affluent countries can contribute towards efforts in this direction, not by cutting down on their purchases of raw materials, but by opening their markets to developing countries wishing to export competitive finished products.

But what is still more important is that the industrialized countries should not intervene to oppose efforts at social transformation in the developing countries. One of the assumptions underlying this summary has been that radical social changes will, in the longer run, imply a reduction of exports of raw materials to the industrialized countries. Preparedness to accept such a development is perhaps the most important contribution that the affluent countries will be able to make towards a more equitable distribution of the world's resources. Preparedness of this kind would in itself imply change, because normally the major industrialized countries have tried to oppose such developments. Here again, Sweden has a tradition to live up to as regards drawing attention to injustices all over the world, and doing what can be done by a small country to support progressive movements in oppressed countries.

Does this mean that Sweden has no narrower and selfish interests to safeguard in the raw materials sector, as one of many egoistic countries? It would be a remarkable thing if we had not. As stated earlier, if our policy aims concerning raw materials were to be codified, they would probably resemble those of most other industrialized countries. It is not possible here to attempt an analysis of the balance between those aims, but a rise in the general level of raw material prices would not necessarily have predominantly negative consequences.

Instead, Sweden's problems may stem from secondary effects, i.e. effects of the measures taken by other industrialized countries in response to tougher conditions in raw materials markets. It is only possible here to outline a few scattered possibilities. Difficulties may arise, if for example, there is a trend towards greater co-operation between developing countries, with the aim of controlling the supply of raw materials, and if the major industrial nations also get organized as a countermeasure. Politically, Sweden could be placed in a difficult situation if a number of consumer clubs were established for the most important raw materials. The International Energy Agency, set up by the industrialized countries in the petroleum sector, has already involved problems of political balance. Difficulties of this kind are bound to be aggravated by the advent of other consumer clubs, with co-ordinated stockpiling policies and more offensive purposes. Clubs of this kind may seem far off today, but co-ordination of economic stocks is a subject which has been discussed in the United States and Japan, and also by the EEC Commission. If the general North-South climate were to become more chilly, Sweden's desire to stand apart from the front lines would not necessarily be appreciated by either developing or industrialized countries.

From an economic viewpoint, there is no lack of potential problems if one cares to look for them. For example, a representative of Rio Tinto Zinc has suggested that the mining companies of the EEC should co-ordinate their purchases of minerals from the developing countries in order to carry more weight, not only in relation to the developing countries themselves but also vis a vis American and Japanese enterprise.[1] Another idea which has been broached, and may come to be seriously considered at some future date, is that of an EEC arrangement whereby the export revenues of the developing countries are stabilized, not through the transfer of free financial resources to compensate for revenue fluctuations, but by guaranteeing the developing countries a certain predictable scope for imports of goods from the EEC, a scope which they pay for with exports whose value may fluctuate, but which ultimately corresponds to their imports. Deficits would then be converted into tied import credits when raw material prices were low, but would be repaid through the excess of exports over imports when demand for raw materials was high in the EEC (and therefore, probably in Sweden as well).

In both cases, especially if the systems thus outlined were introduced by the EEC, Japan and the United States, the effect would be for small countries and small companies outside the blocs to be placed in very constricted market situations. Like other possible arrangements involving tied deliveries of raw materials in return for greater bilateral or regional co-operation, these two examples really serve to show one and the same thing, namely that increasing scale and selectivity in the economic co-operation between states can create difficult situations for small, outside countries with companies that are also small by international standards. It is not certain that developments will move in this direction, but <u>if</u> they do, it is perhaps on this plane that Sweden's raw material problems will exist, rather than in the sense of our being unable to afford higher raw material prices.

(1) P.W. Wright: *Planning Future Raw Material Supply. Long Range Planning, February 1976*

Appendix 3b

<u>Non-renewable raw materials - Summary of a 256pp study by Ernst Hollander (report 311)</u>

Faust: (Goethe)

Deep in thy soil are treasures untold,
an unused, buried richness.
How far the reach of thought, how brave it flowers
still it is too weak to grasp this wealth
and fantasy may fly both high and fast
- for this abundance it will have to halt.
But if you look deep down, as the poet, as the prophet,
you will get a boundless trust in boundlessness.

Hercules : (Hoola Bandoola Band)

Doomsday approaches,
that's what they all say.
Seers and prophets
see the end approach
But if life is Hell for many
and Heaven for a few,
Why go on about Doomsday?
When it'l be the same then as now!

How long will our raw materials last for? Is man's ingenuity so great that we need not worry about physical and biological limits? Or are those limits so narrow that drastic reductions of the material standards in the industrialized countries are the only way to open up the possibilities for living standards worthy of human beings, and especially of the thousand million who are gravely undernourished in the world today?

Many people put the question in these terms when discussing the future. Personally, I believe it is misleading to state the problem in this way, because I am convinced that unequal distribution of incomes and power are important reasons why human creativity today is under-exploited, and often applied to the wrong purposes.

If this is true, the chances of ultimately achieving better living conditions for the entire population of the earth are closely bound up with questions of distribution. This holds good even if the physical and biological limits are not so narrow that they of themselves necessitate drastic reductions of living standards in the industrialized world.

This does not mean that the ecological limits to the type of growth which has characterized the affluent countries during the post-war era lack interest. The answers to questions of this kind are important when, for example, we come to consider whether ecological limits will have a decisive effect in prompting economic and social change, or whether conflicts of other kinds - for example, concerning distribution in the broad sense - will overshadow the ecological issues.

About the Study

This report does not deal with energy raw materials, and it also excludes those raw materials which are termed renewable, the most important of which are foodstuffs and forestry products. The principal raw materials remaining to be discussed are non-energy minerals and the metals extracted from some of those minerals.

Despite these restrictions, our subject is big enough to encompass widely differing opinions.

"We will run out of silver in eleven years' time". "No, the silver will last for 65 million years!"

Those quotes illustrate the spread of assessments about the supplies of non-renewable raw materials. Many people who are worried about the future of the world have difficulty in obtaining and evaluating the data which statements of this kind are based on. One of the purposes of this report, therefore, is to display some of the important pieces of the jigsaw, and a greater number of people will thus be enabled to form their own opinions, and that the discussion will be correspondingly broadened.

The remainder of this summary refers to the content of the report "Non-renewable raw materials", chapter by chapter. Since at the same time the purpose of the review is to bring out facts, problems of method and conclusions, some of the themes will occur in several places.

Minerals - whence and whither?

The paper on which this report is printed contains china clay. Large quantities of sulphur are used in the processes whereby timber is converted into paper. The equipment required for the processing contains iron, carbon, manganese, chromium, nickel and other minerals besides.

What stages have these substances passed through between their discovery and their incorporation into a finished product? Why are many of them very cheap today? Which one's are most important from a supply point of view? The answers to these questions are important to anybody wishing to follow a discussion of the future of minerals.

First, the term "non-renewable". The use of this term does not imply any tacit assessment of the gravity of long-term problems of supply. The term "renewable" might seem to imply that raw materials coming into this category are less of a problem than those which are non-renewable, for things which can be renewed ought not to run out. But in fact the limits to the number of people which the earth can support are more likely to be dictated by the availability of renewable than non-renewable resources.

The point is that some of the non-renewable raw materials occur in vast quantities. Iron and aluminium respectively constitute 6 and 8 per cent of the outer crust of the earth which is accessible to man. The outermost kilometre of the earth's crust has a mass of 10^{18} (billion billion) tons. In practice, therefore, iron and aluminium will never become exhausted so long as there is enough energy available for their extraction.

Other important metals, such as copper, zinc, lead and nickel, however, have average crustal abundancies of between 10 and 100 parts per million. If the contents of these metals in the deposits which are exploited today were as low as this, the metals would be costing hundreds or thousands of dollars per kilo. In certain places, however, they have been very heavily concentrated by geological processes. For example, most deposits of lead contain several per cent of that metal, and this is one of the main reasons why, for a long time now, the price of lead has been only between 20¢ and 70¢ per kilo.

High natural concentration is the result of comparatively rare geological coincidences. Gallium, a metal which has not been concentrated very heavily anywhere by natural means, occurs more commonly in the earth's crust than lead. It costs US dollars 650 per kg, and annual world output is only 15 metric tons, even though this metal has many useful properties. By way of comparison, world output of lead amounts to more than 4 million tons a year. Thus the proportion of the total deposits of a metal occurring in heavy concentrations is a very important factor in determining how much of that metal will be available to human society in the future.

But mineral supply is not only a question of metals. Minerals used on account of properties other than metallic content are called industrial minerals. Because many of them occur very abundantly, they are often cheap. Their importance has consequently grown, and can be expected to keep on doing so.

The basic features of world mineral production - a flash picture of today's key problems

Historically speaking, the strategic role of metals has been tremendous. For example, Sweden's career as a great power in the seventeenth century hinged in many important respects on revenues from the Falun copper mines, which at that time accounted for two-thirds of world output of copper.

Although the strategic role of metals is presumably smaller today, metals still play a fundamental part in the economies of many countries. New facilities for extracting and refining many of the most important metals, on the scale required to minimize costs, involve a capital outlay of between 500 and 1,500 million dollars. Most small countries have difficulty in financing such enormous investments internally, added to which the resultant facilities are so large that nearly all metal-producing countries have to export considerable proportions of their output. Taken together, these factors can have serious repercussions on the national independence of small mineral-producing countries.

A lot of energy is need for the extraction and processing of minerals and, above all, in order to obtain pure metals, ready-to-use building materials etc from them. These stages of mineral processing (together with the transport operations involved) are estimated to account for at least 20 per cent of the total energy used in the world. It follows that an energy shortage (if there is one) will profoundly influence the cost of mineral-based products.

Thirty mining groups, which in many cases are closely allied with each other, account for around 45 per cent of the western world's output of non-fuel minerals.

Through ownership and other forms of control, these groups often exert paramount influence on the other 120 mining enterprises which account for most of the remaining 55 per cent of world output. The implications of this commercial concentration are not analysed very closely in the present report, but it is clear that knowledge concerning the value of different mineral deposits, and opportunities of exploiting them, are confined to a small number of people. This general statement is only marginally modified by the knowledge that there is a substantial number of small mines in the world partly run outside the direct control of the big companies.

Developing countries with market economies account for no more than about a quarter of the world's production of (non-fuel) minerals, which is very small by comparison with their share of the world's land area (50 per cent) or their share of world population (48 per cent). Since, however, they only consume 6 per cent of world mineral production, they are still the greatest net exporters; the planned economies have only limited surpluses in relation to the rest of the world. But the developing countries have no "non-fuel-mineral-power" comparable to the "oil power". Before oil prices went up, the total value of non-fuel mineral production was only half that of oil, and at the same time the developing countries had much smaller shares of the market. What is more, the ruling strata in most developing countries that produce large quantities of non-fuel minerals are closely allied with the ruling strata of the major purchasing countries.

Serious mining investments normally require planning between twenty and thirty years ahead. For this reason, long-term political stability is in most cases a precondition of new investments in a country. This is an important reason for the post-war stagnation of the Third World's total share of mining output.[1] The transnational enterprises have concentrated their new investments in certain industrialized countries (Australia, Canada, South Africa). In the few developing countries where investments have been made by these enterprises, long-term political stability and big profits have often been guaranteed by widespread oppression.

There is a risk that future increases in developing countries' output will be based on the same preconditions. But it is also possible that developing country governments will derive their power from widespread popular support and will strive to increase mineral exports in order to finance economic and social development. It is therefore important to pay attention to the political framework when deciding one's standpoint concerning more intensive exploitation of the great untapped mineral resources of the developing countries.

The international and domestic physical environmental problems of mineral production are more serious than those affecting most other industry sectors. The ecological disruption problems may be aggravated when the exploitation of poorer deposits leads to the involvement of larger land areas. Intensified conservation laws operate in the other direction, but many of the environmental hazards are so complex that they are difficult to record statistically, and therefore difficult to legislate about.

In Sweden, as in many other countries, mines head the statistics of working days lost on account of accidents. These accidents are declining in Swedish mines, but the dangers in working conditions also include many long-term hazards which are not reflected in the statistics that are kept today.

Far heavier expenditure will probably be required in order to improve the ecological

(1) The stagnation of the share still implies a growth in Third World mining output in absolute terms. However this growth has been smaller than what would have been considered reasonable in relation to known geological facts about these regions.

disruption to such an extent that the inhabitants of areas affected by mineral extraction will voluntarily accept that production starts, and in order to improve working conditions to such an extent that people will freely choose to work in mines even when there are other acceptable forms of employment available.

Consumption Forecasts: Uncertain Yesterday and Perhaps even less Certain Today

Very large economic resources are usually needed in order to make good mineral forecasts.

Institutions and enterprises which command a full view of the long and complex chain of operations, from the minerals in the ground to the finished products leaving the factory, have still very great problems to contend with, because good mineral forecasts require a picture of the growth of the entire world economy, the development of mineral consumption being very closely bound up with economic growth as a whole.

Forecasting is a particularly hazardous business today, since earlier post-war patterns of global economic development most probably have come to an end.

There are, however, some important conclusions which can be drawn with the aid of forecasts. Given a world economy working on essentially the same lines as at present, consumption can be expected to rise very steeply during the next twenty-five years, so that by the year 2000 consumption of many minerals will be twice or four times what it is now.

We have presented some hypothetical calculations meant to show what might happen if different rules came to apply, i.e. if developments were set in train with the object of bringing world per capita consumption of minerals up to the level of present-day per capita consumption in the affluent countries.

The conclusion drawn from this calculation is encouraging in a sense, because it indicates that development aimed at global justice would not require more minerals than "spontaneous development". Thus if resources are sufficient for "spontaneous" development, they would also be sufficient for the income equalization on a high per capita levels.

Long-term Availability. Some Methods of Measurement and their Limitations.

There are a number of rather complex concepts which have to be mastered before one can proceed to assess our possibilities of coping with "spontaneous development of demand". These concepts are used to describe the volumes of different minerals which will be available in the long run.

One reason why these concepts are so complicated is that they are used to describe a reality which we do not know very much about. Mineral prospecting is expensive. It is very hard to tell what techniques may come to be used in the future for the extraction and refining of minerals. In addition, most of the work done to explore the world's resources is governed by considerations of business economics. Normally this work forms part of the long-term planning of private or state-owned enterprises. Investments in prospecting often have to meet the same profitability requirements as investments of other kinds. In practice, this means that a company will seldom acquire mineral reserves corresponding to more than about twenty years' output. Reserves only include those mineral deposits which have been carefully investigated and which are presumed to be worth exploiting at present-day mineral prices.

Attempts are frequently made to add up all the reserves in the world. It is not surprising that, when these figures are used to gauge the amount of minerals available in the long term, the conclusion is reached that minerals will be exhausted in about twenty years' time. One should thus be cautious with conclusions based on figures concerning reserves, all the more so as data on this subject represent strategic and expensive information which can be distorted for many reasons.

Reserves are one of the concepts present. The others are resources and the total content of the earth's crust. The two statements about silver quoted at the beginning of this summary represent the grossest misapplications of the terms 'reserves' and "total content of the earth's crust" respectively.

Figures concerning reserves are intended to show the volume of minerals that will be available when more prospecting has been done and if presently unremunerative deposits are exploited. More often than not, these figures are even less reliable than the figures for reserves, because in order to be able to calculate on the basis of such figures, one must disregard technological progress completely (or else try to foresee the course of technological development during the twenty years for which existing reserves may last).

The total content of a certain metal in the earth's crust is sometimes referred to for demagogic purposes by certain growth enthusiasts, but the type of extraction thus implied would involve quite preposterous energy costs per ton of metal produced. An account of the historical development of reserves does however show that reserves have so far kept pace with output. This is because prospecting and technological progress have hitherto been powerful instruments in counteracting the depletion of resources. Consideration is also given to a third instrument used for maintaining reserves at the desired level, namely price rises. But there are only a few metals whose prices have shown a long-term increase. The overall impression left by this inquiry is that it is very difficult to gauge the state of mineral assets fifty or a hundred years ahead, because historical analogies do not have very much to tell us in such a perspective.

Shortages of Individual Metals?

So what can we do? Our investigation has led to the conclusion that the physical limits are remote for most metals. For practical purposes, however, some metals would "run out" if the 3 per cent annual rise in consumption were to continue for another 75 or 100 years. On the other hand, if events were to take this course, by the middle of the next century there would be enormous quantities of recoverable scrap metal. The whole situation is quite improbable, however, because the metals in question would presumably be replaced by other materials for most purposes long before the situation could arise.

The physical limits seem quite definitely to be so remote that they need not of themselves prevent the developing countries from attaining the same average per capita consumption as exists in the affluent countries today.

Although the future for metals does not look at all ominous in the narrow context of availability, it is interesting to point out the metals whose growth rate can be expected to decline because of scarcity. A relative decline in the consumption of these metals may result from their first becoming more expensive, but the very risk of a shortage may be sufficient to cause them to be increasingly replaced by other materials. Processes of this kind, involving the substitution of one type of material for another, can create adjustment problems in different sectors and communitites, and entire countries may also be affected. If the return on a scarce metal is boosted by rising prices, producing countries may benefit for a short period, but in the longer term they will be obliged to change the focus of their production, and this can often cause problems, particularly to developing countries, because their economic structures are often more rigid than those of the industrialized countries.

If we confine our attention to the remainder of the twentieth century, a list of metals which may come to be in shorter supply than they are today would, in the first place, include tin and silver, though tungsten and possibly copper may also cause certain problems within the same period. Looking further ahead, lead and zinc should be added to the list, together with gold and the platinum group metals.

However, there is a great deal of uncertainty involved in lists of this kind. In certain situations, the people in control of information may wish to create an image of scarcity, while in other situations they may wish to convev the impression of boundless resources, and they have a very good chance of making both versions appear as accepted truth.

Shortages of industrial minerals?

Hardly any industrial minerals are expected to run short during the remaining years of this century. Looking very far ahead - a matter of several hundred years - supplies of phosphorous may present a problem. If so, this would be very serious, because phosphorus is a fundamental and irreplaceable ingredient of commercial fertilizer, and an abundant supply of commercial fertilizer is one of the most important technical prerequisites of a steep rise in food production.

Taken as a group, however, industrial minerals present more of a promise than a threat. Their abundant availability, coupled with their energy-saving virtues, may cause industrial minerals to become substitutes for metals to an increasing extent.

Shortage of Non-Fuel Mineral Products

A frequently quoted American report already mentioned in this book[(1)] has shown that mineral extraction costs have been falling steadily since the end of the nineteenth century. These findings have sometimes been taken as evidence to show that rising consumption of raw materials does not lead to shortages, because technological progress more than offsets the effects of the depletion of the best deposits. This is presumably reading too much into the findings, but it is clear that the measurable costs of mineral extraction have been on the decline so far. It is important, however, to emphasize that the costs referred to are measurable costs, because the extraction and processing of minerals involve a series of environmental costs whose long-term effects are not easily sized up.

We have tried to outline, rhapsodically, some conceivable developments of the global supply situation through a fairly mechanical arithmetical exercise covering the next twenty-five or fifty years. The exercise is based on a world economy operating more or less on the same lines as at present, and it is characterized by quite pessimistic assumptions about the future conditions of mineral extraction. In other words, the exercise is intended to bring out possible problems. Thus it is assumed that the cost of minerals which may run short will rise steeply, that energy will be more expensive, and that heavy anti-pollution investments will be needed. No allowance is made for the possibility of resource-saving techniques or material-saving consumption. Even so, the total share of world GNP applied to the production of finished metals and industrial minerals does not come to more than 7 per cent in the exercise outlined above, as against 3 per cent today. Most of the increased cost can be attributed to rising energy and environmental costs.

(1) H.J. Barnett and C. Morse, "Scarcity and Growth - the Economics of Natural Resource Availability, Resources for the future, 1963.

Many countries would obviously run into serious problems faced with the adjustments demanded by such an increase in the costs for the production of metals and finished minerals, but these adjustments per se would hardly be likely to jeopardize the survival of the societies concerned. The parallel or rising oil prices can furnish some guidance concerning the nature of the effects, but the comparison has its limitations because oil prices rose very quickly, whereas the changes referred to here would presumably be spread out over quite a considerable period of time.

The first conclusion drawn from the exercise is somewhat counter-intuitive. Given a reasonable amount of time in which to adjust, the non-fuel mineral sector should be quite capable of bearing the additional costs caused by societal demands on the sector as a consequence of problems concerning the working conditions and ecological disruptions, difficulties of social planning due to the extreme fluctuations of profitability in this sector, its localization, etc. If demands of this kind were only to be made on firms in the mineral sector of an individual country, those firms would probably be put out of business by international competition. International initiatives are therefore required to solve these problems. Waiting for such initiatives to become successful however, should not be allowed to stand in the way of steps to deal with the problems. The countries which act in time to create the new political forms, whereby the burdens of adjustment can be fairly distributed, and which make plans for the new technology which is going to be needed, will have a good chance of gaining by their actions in the long run.

Perhaps an even more important conclusion is connected with the way in which costs are presumed to increase. If it is a correct assumption that energy costs and environmental costs will feature largest, the most important ecological arguments for saving on minerals will not be those connected with availability in the narrow sense. Instead, they will relate to the energy and the environmental values that can be saved by restraining the rise in consumption. In this case, one can argue forcefully for better management of non-fuel mineral-based products without contending that they are in danger of running out.

What do we mean by better managment of non-fuel mineral-based products? Many different things, presumably, but the following appear to be the most important of them.

The growth of consumption should be concentrated within the countries whose needs are greatest, i.e. the developing countries. As has already been stated, there are no physical impediments to prevent the developing countries from achieving the same level of average mineral consumption as already applies in the industrialized countries today. The impediments are of a different kind.

Firstly, fundamental changes will have to be made in the economic, social and political organization of the developing countries. These changes include a radical redistribution and equalization of power and incomes. When the poorest inhabitants of the developing countries obtain provision for their basic needs, they will be able to make a substantially greater contribution to production. Reduced consumption of luxuries will mean that certain resources can be liberated for other uses and, above all, that the foundations of social solidarity will be laid. In this way, redistributions will enable the developing countries to set aside a much larger portion of their resources for saving, which is an essential precondition of any seriously intended agricultural and industrial development.

Secondly, the industrialized countries will have to stop supporting regimes opposing this type of development.

Thirdly, the industrialized countries will have to transfer considerable resources to the developing countries. (This is where some of the changes included in the expression, a New International Economic Order, come into the picture).

A programme of this kind may seem utopian, and it is unlikely to be feasible unless the peoples of the developing countries fight hard to achieve it. Initially, therefore, it will demand great sacrifices on the part of the large groups of people in the developing countries (and probably by large groups of people in the industrialized countries too). Nonetheless, development in this direction will probably be necessary in order for the fifty per cent of the earth's population living in the developing countries to achieve decent living conditions in the long run.

Also in the long run, a reduction of the income gaps between industrialized and developing countries can probably benefit industrialized regions like Europe and Japan, which are poorly endowed with raw materials. Unless these gaps are reduced sufficiently for a North-South confrontation to be averted, Europe and Japan may become heavily dependent on countries like the United States, Canada, the Soviet Union and Australia, where supplies of raw materials are more plentiful.

Obviously, the changes which will have to take place in order for global management of mineral resources to be improved are not so much concerned with questions directly relating to minerals. This impression is supported by a discussion of conditions in the <u>industrialized</u> countries.

Although utter material poverty still occurs in many of the industrialized countries, the greatest problems confronting most of them are of a different character. Those problems are concerned with the lack of certain fundamental qualities of life.

An overview of how our own daily work fits into a broader pattern, and the possibility of influencing its long-term direction, are two such qualities of life. Possibilities of influencing the environment are another example. For the sake of comprehensive personal development, people need to be able to alternate between different duties - intellectual tasks, manual activities, and duties relating to the care of children and old people. Quality of life can also include such things as a reduction of stress and deeper sense of community with other people.

There is a great deal to suggest that extensive areas of consumption of goods in the industrialized countries are a kind of substitute for these qualities of life. If, therefore, societies can be created which are distinguished by the above-mentioned qualities of life, the standard of living in the broad sense would probably rise even if there were a decline in material consumption.

A process towards deeper democracy aimed (among other things) at making full use of human creative capacity would probably also have vital effects on technological developments. If more people came to play an active part in the development of new techniques, some of the negative aspects of technology, which frighten many people today, could possibly be rectified. This in turn would influence the conditions of future raw material supply.

The above remark illustrate that each person's conception of man's nature also determins the assessment of how problems concerning raw materials should be with. To anybody who believes that narrow self-interest is the only conceivable driving force of human action, the type of development outlined here is bound to appear utopian.

The material in this report does not suggest that supplies or shortages of non-fuel minerals will <u>necessitate</u> extensive social changes. Whether the supply of energy, or environmental menaces in the broad sense, are of such a coercive nature is not a question which can be answered on the strength of the data put forward here. It is my belief, however, that international justice and qualities of life, such as those discussed above, are intrinsically valuable and that work for their attainment does not need to be justified in terms of the physical limits to our existence.

Ernst Hollander

Appendix 3c

The substitution of resources[1]

Summary of 240 p. report. (Report 309)

Editor and main author: Uno Svedin. Contributions by Tommy Bengtsson (Chapter on Economic History), Karl Jungenfelt (Chapter on Economic Theory) and Goran Grimvall(Chapter on Technology)

Introduction

What is Substitution

Perhaps the first thing which comes to mind concerning the substitution of raw materials is the forced changes from one raw materials to another that must take place when, for example, a country is blockaded. During the last war, cars in Sweden had to be fuelled with producer gas instead of petrol, which could no longer be imported. Ersatz coffee was manufactured because of lack of the natural raw material, and the same thing happened with rubber.

In the normal run of things, however, the substitution of one raw material for another is not caused by a basic resource completely disappearing, and having to be replaced as far as possible with something else. More commonly, the substitution process is a reflection of ongoing technological progress. The change from flint implements to bronze was due to the superior efficiency obtained with metal, and not any shortage of flint. Tools became better still when a technique was devised of producing and using an even harder material, namely iron. This process goes on today. The time is long since past when timber was the main material used for bridge building. Since then we have learned to master other materials whose superior properties make for a sounder construction.

In the heated debate which flares up from time to time concerning the survival of humanity and, in the same connection, concerning a global collapse of resources, opportunities of substitution have sometimes been referred to as a phenomenon to

(1) The text in this appendix was originally used as a short introduction to the report on Substitution. Thus it should not be regarded as an independent source of information but rather as a short indicator of type of material which is available in the (Swedish) report.

counteract the scarcity of resources. This has been expressed in the following terms by the Swedish economist Assar Lindbeck(1):

"The fact that we are unable today to specify exactly which goods and which methods of production will predominate in twenty-five or a hundred years' time, is no reason for jettisoning the knowledge we actually possess concerning the way in which production is, 'in principle', constantly being adjusted".

The American economist Robert B Solow has observed that "the two forces standing between ourselves and a definitive shortage of resources are technological progress and opportunities of substitution".(2) In the following discussion, these two forces will be regarded as closely allied but essentially different phenomena. Technological progress often makes substitution possible, but does not of itself amount to substitution. Substitution can be a manifestation of technological progress (e.g. a new and cheaper process for the production of a material) but it need not be, because there are other possible causes. Changes not necessarily connected with technological progress can influence the choice of commodity, by transforming the structure of prices. Changing habits may also increase the demand for resources of new kinds, and reduce the demand for others.

Degrees of interchangeability

Purely scientifically speaking, two substances can never be wholly interchangeable. Elements, after all, have different properties. But there are quite a number of technological applications which only utilize certain of the properties of elements, and elements may be similar enough as regards those properties to be readily interchangeable. Thus the question of interchangeability always has to be viewed in relation to the use made of an element in a particular situation.

Quite often the set of properties aimed at in a certain technological construction can be achieved through a combination of different elements. In this case, interchangeability means the possibility of obtaining the technical function by using a new combination.

This is possible where certain types of function are concerned. It is less easy with others, particularly with those connected with biological processes, where requirements are often very closely defined and the direct opportunities of variation are very limited. To take one example, phosphorus is used in artificial fertilizers. It is absolutely essential for this purpose and is not interchangeable in the production of grain. And yet there is still a certain scope for manoeuvring in cases of this kind, because the non-interchangeable element can be obtained in a variety of ways. Phosphorus can be obtained from minerals of different kinds or it can be recovered from bone material. Plants need nitrogen, but this can be obtained either by utilizing nitrogenous saltpetre deposits, or else by binding atmospheric nitrogen by technological or biological means.

Even if two elements are interchangeable for a certain function, the extent of their interchangeability can vary. German Ersatz coffee during the last war was pre-eminently a substitute, but it was inferior in quality to the natural raw material which it replaced. Synthetic fibres were of inferior quality to natural fibres when they first began to be marketed. The same was true of artificial rubber compared with natural rubber. In time, however, the quality of the substitute was improved. In general terms, therefore, interchangeability may be said to exist on a sliding scale.

(1) Forskning och Framsteg ("Research and Progress"), no. 5/1975.

(2) Ekonomisk Debatt, no. 8/1974, p. 492

Assessments vary concerning what is or is not interchangeable. The sceptics maintain that we lack the technological "elasticity" which will be needed to adapt society's future material systems to the raw materials which will be available in large quantities. The debate on this subject has to some extent assumed the guise of a contest between the economic and the natural science approach. All the more reason, then, to examine the problem of substitution from as many different vantage points as possible.

Different types and levels of substitution

Substitution does not only occur between one material and another, but also between a material and energy, as for example when efforts are made to insulate buildings (consumption of building materials) to bring down the amount of energy required for heating purposes (energy saving). Material-material and material-energy substitutions are two examples of what we will call types of substitution. But a change can also occur through the replacement of an entire technological system (e.g. the substitution of electrical power for the piston engine in motor vehicles). This is termed substitution at a higher level. Of course, it also has repercussions on materials. Substitution can also be brought about by changes in demand, i.e. on the consumer side, due for example to changing values. This is an example of another level of substitution.

Great economic flexibility - some of the findings

In the review undertaken on economic premises, it is found that the material system constitutiing our resource base is highly flexible. Substitutions are made not only between different materials, but also between energy, labour and capital. During the past hundred years, technological advances have helped to secure steady reductions in the real costs of producing a unit of output (one tone of ore, for example). (Rising costs would not have been surprising, considering that progressively inferior resources have had to be utilized, e.g. less fertile soils and lower grade ore deposits).

There is nothing to guarantee the automatic continuation of this pattern of development but, when analysing the resource picture of the future, it would be a patent error to underrate the flexibility possessed by the system of materials in combination with technological progress. A model of the future in which the use of natural resources is characterized by complete rigidity cannot claim to give a likely description of our future situation.

Present-day pricing mechanism not good enough

On the other hand, there are many arguments indicating that the present-day economic system does not provide all the mechanisms which can be needed in order for the long-term supply of resources to be provided as efficiently as possible. This applies particularly to the inability of the price system to give an adequate reflection of a possible long-term shortage. It is not easy to come up with alternative solutions to this problem. The resource situation of the future will depend on the technological development which then has materialized. This is unknown to us today, even in very general terms, and it therefore constitutes a genuine source of uncertainty. Short-term choices between different substitutes are often based on a strict comparison of prices. The decision as to what is to be included in the comparison of costs and what should be taken as given (e.g. investments already made and the way in which they are to be written off) does a great deal to influence this choice, and does the way in which various side-effects are included in the costs.

Technical interchangeability - a question of new functional solutions

The argument concerning continuing possibilities of economic development through the gradual replacement of raw materials which become increasingly scarce, is based on the presupposition that there is no specific substance which is indispensable in the production of an absolutely essential commodity. It can be observed that, theoretically speaking, there are logistical hazards connected with technology based on "key materials", such as tin, lead, copper, tungsten, silver, platinum, etc. But there is more to this problem than the availability of substitutes for these particular substances in present-day products. The development of materials hinges on other things besides deposits of a particular element. For example, there are practically no instances of irreplaceable metals being used as alloying element. The important thing is the function sustained by the technological solution. The further we try to look ahead, the greater is the importance attaching to substitutions at a higher level in the form of solutions based on completely new principles. If instead of carrying people from one place to another, we transmit information, using for example an electronic tele-viewer system which enables people to meet without travelling, the entire transport sector will be affected. If we develop an efficient public transport system instead of resorting to private motoring, this may conceivably affect both the total consumption of resources and the type of resources employed.

Shift in the pressure on resources

Every new solution causes the pressure on resources to be shifted from one type of resources to another. New solutions of a resource-saving type may demand heavier input of manual labour or capital. Waiting periods for transport services may be lengthened, and private convenience may be reduced for the sake of certain alternative transport solutions. This is another type of cost which has to be taken into account. The environmental consequences of new solutions (not only in connection with production, but also in connection with use and waste products) must also be borne in mind.

On the technological side, greater efforts may be required of the designer, who must make a more exact calculation of the consumption of certain limited materials. The education and experience required for the utilization of a new material may limit the potential impact of certain substitutions.

Are substitution processes similar?

Substitution processes can differ markedly in character, depending on whether they are sudden changes born of a crisis, or a response to more profound and long-term tendencies in the world at large. Many of the substitutes developed during the last war did not survive very long after wartime restrictions were abolished in 1945. Either they were of inadequate quality, or else they were too expensive compared with the products afforded by the alternative methods which became available once more after the war. Substitutions established in less critical circumstances often take quite a considerable time to gain ground. The time needed for the new material, compared with its rivals, to progress from a ten per cent to a ninety per cent share of the market has been calculated to be about fifty years - if present trends continue - in the case of plastic as against real leather, or synthetic as against natural fibre. In some cases, on the other hand, the transition period can be as short as ten years. This happened when synthetic detergents took the place of natural soap in the USA in about 1950. The interesting thing about many substitution processes is that they follow the same essential course but at different speeds. Once the process has started, it often continues until the new material has entirely, or almost entirely, supplanted the old one. The change proceeds slowly to begin with, due among other things to the bottleneck caused by investments in plant and training for the old material, as well as the lack of knowledge in the market concerning the new material. Eventually the substitution process gathers momentum, and it continues until a great deal of the market has been taken over by the new solution. The pace then diminishes again, and the last few per cent of the substitution may take a considerable length of time to accomplish.

Often the underlying circumstances do not change very much during the substitution process. It would be a mistake, however, to conclude from this that, once a substitution process has begun, it continues irrevocably and permanently in one direction only. If there is a change in the circumstances which helped to bring about the process, substitution may occur in the opposite direction. We have a historical instance of this kind in the transition from wood to coal and then to oil, for purposes of fuel. The plans for energy forests (i.e. wood), which are now being discussed, as well as current conjectures regarding the use of carbon powder to fuel motor vehicles, show that new technical circumstances can lead to the reversal of substitution processes. However, it often takes a long time for these circumstances to change.

<u>Substitution- the long-term method of overcoming a scarcity of resources</u>

What are we to expect concerning the substitutions which must be effected sooner or later to prevent the exhaustion of more limited resources, and even leading to a general scarcity of resources? Two American researchers H.E. Goeller and A.M. Weinberg, have outlined a course of development covering two centuries and implying a transition from the materials we use today to materials which are available in large, practically inexhaustible, quantities. Development in this direction is conditional on a technical and environmental solution being found to the energy problem.

Whatever the substitutions made, either in the short term or in the long term, they are always bound to have indirect consequences. Many developing countries today depend for most of their export revenues on one or two raw materials. Their substitution by other materials can spell economic disaster for these countries. Similarly, in more industrialized countries, substitutions can have regional consequences, not least for employment. Substitutions can also require the use of new kinds of raw materials. This can have environmental repercussions on completely new geographical areas. The use of unexploited wetland for the energy forests now being planned, as well as demands on uranium deposits (which in the early years of this century would not even have been classed as resources), are examples of emergent conflicts of aims due, in the ultimate analysis, to substitutions (in this case, between different raw materials for the production of energy). In a more general context, substitutions can transform social conditions. The outsting of flax (linen) and wool by cotton during the 19th century transformed the conditions of cottage-based industry. The raw material was no longer produced locally, it was imported, and this had a bearing on business finance. Now that it was no longer possible for the single individual to buy raw materials, the decision-making process became more centralized and greater use was made of wage labour.

The measures which have to be taken to bring about a favourable course of development include a whole series of efforts, from learning more about substitution processes to investing in substitution preparedness, and efforts in production processes are especially important in the shorter run. The evolution of new techniques calls for research and development by government research bodies and private enterprise. The State must also accept responsibility for co-ordination and make sure that social conditions are favourable enough for an adequate lead to be maintained.

In the longer-term, co-ordinated efforts will probably be needed at the international level. Ultimately, however, changes in long-term values will probably be the most important factor. These values will decide the extent to which different products and services are in demand and, therefore, the relative importance of rival substitutes.

Appendix 4 — Statistics

Table 1. World production of selected agricultural commodities between 1969-1971 and projections up till 1990

	Consumption	Extrapolated demand			Total increase		Rate of increase	
	1969-70	1980	1985	1990	1985/70	1990/70	1970/85	1970/90
	Million metric tons				Percent		Percent per year average	
All cereals (1)	1207	1538	1725	1910	42,9	58,3	2,4	2,3
Wheat (1)	332	404	447	490	43,8	47,6	2,0	2,0
Rice (paddy)	310	400	447	493	44,3	59,1	2,5	2,3
Feeding Grain (1)	565	734	831	927	46,9	64,0	2,6	2,5
Roots	279	321	342	361	22,3	29,2	1,4	1,3
Sugat (centrifugated)	70	92	107	126	54,1	80,9	2,9	3,0
Sugar (non-centrifugated)	12	15	17	19	42,1	54,0	2,4	2,2
Podded plants, nuts, oil seeds	52	69	79	91	51,2	73,5	2,8	2,8
Vegetables	223	290	330	374	48,0	67,7	2,6	2,6
Fruit	158	214	250	290	58,2	83,5	3,1	3,1
Meat (2)	107	144	168	197	57,3	84,7	3,1	3,1
Beef and Veal	39	51	60	90	54,7	80,1	3,0	3,0
Mutton and lamb	7	10	12	14	69,1	104,5	3,6	3,6
Pigmeat	36	46	53	61	48,3	71,4	2,7	2,7
Poultry	16	24	29	36	83,5	125,4	4,1	4,1
Eggs	19	25	29	34	50,8	75,1	2,8	2,8
Fish	41	57	68	81	64,6	96,3	3,4	3,4
Whole milk including butter (3)	389	476	532	597	36,9	53,3	2,1	2,2
Skim milk	(40)	(50)	(58)	(66)	(44,4)	(65,5)	(2,5)	(2,6)
Cheese	9	12	14	16	51,7	74,4	2,8	2,8
Fats and oils (4)	33	43	49	57	47,9	70,8	2,6	2,7
Butter (fat content)	5	6	7	8	33,9	49,3	2,0	2,0
Vegetable oils	22	30	35	41	55,1	81,9	3,0	3,0

(1) *Including demand for fodder and for other purposes than food, expressed in primary products*
(2) *Including waste*
(3) *Including milk products, expressed in milk*
(4) *Including animal fats*

Sources: For 1969-71, i) OECD - Food Consumption Statistics
ii) FAO - Supply Utilization Accounts - Statistics Division
For 1980, 1985 and 1990: FAO

Table 2A. World primary mineral demand in 1974 and projected average yearly growth in demand up to the year 2000 according to US Bureau of Mines

Commodity	Units (metric tonnes)	Consumption in 1974	Growth in demand (%/year)
STEEL INDUSTRY METALS			
Iron	million tons	510	2,7
The five most important ferro-alloys			
Manganese	thousand tons	9300	3,1
Chromium	" "	2450	2,6
Nickel	" "	700	2,8
Molybdenum	" "	91	4,2
Tungsten	" "	39	2,8
Some other ferro-alloys			
Cobolt	" "	31	2,7
Columbium	" "	13	4,5
Tantalum	" "	1,4	3,8
Vanadium	" "	25	4,6
NON-FERROUS METALS			
Base metals			
Copper	" "	7300	3,9
Zinc	" "	5800	2,6
Lead	" "	3700	2,3
Tin	" "	230	1,6
Light metals			
Aluminium	" "	15000	5,4
Magnesium	" "	279	4,4
Titanium	" "	53,5	3,5
Beryllium	" "	0,5	3,8
Precious metal			
Gold	tons	930	2,4
Silver	"	13000	1,7
Platinum	"	84	2,3
Rhodium	"	5	2,2
Palladium	"	51,3	4,4

Source - see p. 185

Commodity	Units (metric tonnes)	Consumption in 1974	Growth in demand (%/year)
Some other non-ferrous metals			
Antimony	thousand tons	70	2,4
Cadmium	" "	17	3,0
Mercury	" "	8,1	0,5
Bismuth	" "	2,7	3,6
OTHER METALS AND SEMICONDUCTORS			
Silicon	" "	1800	3,8
Selenium	" "	1,2	4,1
Arsenic	" "	45	0,4
Germanium	tons	70	2,3
Hafnium	"	59	3,3
Indium	"	56	2,4
Gallium	"	12	5,4
Rhonium	"	3,2	3,0
Scandium	kg	12	4,3
Tellurium	tons	220	1,9
Tallium	"	9,5	0,0
Rare earths			
Yttrium	thousand tons	24	4,0
Zirkonium	" "	6,3	7,3

Table 2B. World reserves and resources of metals 1974 divided by world primary cummulative primary demand 1974-2000

	Reserves 1974	Units(1) (million tonnes)	Reserves / Cumulative demand	=Resources / Cumulative demand(2)
STEEL INDUSTRY METALS				
Iron	91	billion tons	4,5	
<u>The five most important ferro alloys</u>				
Manganese	1826	million tons	4,9	
Chromium	523	" "	5,7	
Nickel	54	" "	2,1	4,2
Molybdenum	5,9	" "	1,4	
Tungsten	1,8	" "	1,2	3,5
<u>Some other ferro alloys</u>				
Cobolt	2,45	" "	2,1	
Columbium	10,9	" "	10	
Tantalum	0,068	" "	1,1	
Vanadium	9,7	" "	7,5	
NON-FERROUS METALS				
<u>Base metals</u>				
Copper	408	" "	1,3	5,8
Zinc	236	" "	1,1	
Lead	150	" "	1,2	2,4
Tin	10,3	" "	1,3	2,7
<u>Light metals</u>				
Aluminium (only bauxite)	3480	" "	4,0	
Magnesium	sufficient		>10	
Titanium	340	" "	4,4	
Beryllium	380	thousand tons	>10	
<u>Precious metals</u>				
Gold	41,1	" "	1,3	1,8
Silver	187	" "	0,4	1,7
Platinum	9,24	" "	3,1	6,9
Rhodium	0,53	" "	2,8	8,7
Palladium	6,03	" "	2,4	>10

	Reserves 1974	Units(1) (million tonnes)		Reserves / Cumulative demand	=Resources / Cumulative demand(2)
OTHER NON-FERROUS METALS					
Antimony	4140	thousand	tons	1,6	2,0
Cadmium	753	"	"	1,1	
Mercury	170	"	"	0,7	2,7
Bismuth	59	"	"	0,5	1,1
OTHER METALS AND SEMICONDUCTORS					
conductors					
Arsenic	3460	"	"	2,8	
Hafnium	sufficient			>10	
Indium	1,5	"	"	0,8	1,7
Gallium	sufficient				
Rhenium	2,4	"	"	>10	
Scandium	sufficient				
Tellurium	40	"	"	5,4	
Tallium	0,3	"	"	1,2	
Rare earths					
Yttrium	7046	"	"	6,3	
Zirconium	20000	"	"	2,0	
Selenium	168	"	"	3,0	
Germanium	1,8	"	"	0,7	1,5

(1) All figures refer to metal content

(2) Generally only figures below 5 given specifically. For copper and for platinum, Rhodium and palladium also non-discovered resources have been included. This is the reason why these figures are given explicitly despite the fact that they are larger than 5.

Sources: Data from USBM 1975 table 12 and 13 (page 23 and 33). The classification of commodities is our own.

Table 3. The fraction of the world's production and consumption of certain minerals held in 1973 by the industrialized, the developing and the planned economy countries

	Industrialized countries		Developing countries		Planned economy countries	
	Percentage of the world's					
	Production	Consumption	Production	Consumption	Production	Consumption
Energy raw materials	32,1	63,7	40,4	10,2	27,5	26,1
Bauxite	37,8	76,6	48,4	5,4	13,8	18,0
Copper	44,6	74,6	35,9	4,6	19,5	20,8
Tin	8,5	74,5	75,1	6,2	16,4	19,3
Iron and steel	40,3	61,6	25,1	8,0	36,4	30,4
Lead	51,5	67,0	21,7	7,5	26,8	25,5
Zinc	55,8	72,4	22,5	7,9	21,7	19,7
Phosphate	44,2	57,8	30,7	14,1	25,1	28,1
Potash	56,7	55,3	3,7	9,3	39,6	35,4
Population	19,8		48,3		31,9	

Source: Annales des Mines, December 1975

Table 4. The change of the prize level at an increase of raw material costs of 100% and 200% for different sectors

	Sector	100%	200%
1.	Agriculture, fishery	1.06	1.11
2.	Forestry	0.97	0.94
3.	Extractive industry	1.08	1.16
4.	Import protected food industry	1.05	1.10
5.	Food industry exposed to foreign competition	1.26	1.49
6.	Drink, beverage, tobacco	0.99	0.96
7.	Textile and clothing industry	1.09	1.17
8.	Wood, pulp and paper	0.98	0.96
9.	Graphic industry	0.97	0.94
10.	Rubber industry	1.15	1.29
11.	Chemical industry	1.02	1.03
12.	Petrol and coal	1.02	1.04
13.	Soil and stone	0.98	0.97
14.	Iron, steel and metal	1.40	1.77
15.	Engineering industry (excluding shipyards)	1.12	1.23
16.	Shipyards	1.20	1.38
17.	Other manufacturing industries	1.05	1.10
18.	Electric, gas, heat and water	0.97	0.93
19.	Building industry	1.03	1.05
20.	Trade	0.97	0.93
21.	Communications	0.98	0.96
22.	Building administration	0.97	0.94
23.	Service sector	0.98	0.96

Source: H Horn af Tantzien och U Karlstrom, Handelshogskolan i Stockholm

Table 5. Annual waste in Sweden (first years of the 1970's) (in thousand tons per year

Domestic, office and trade waste		2,500
Waste from industry (general)		3,500
Waste from building industry		500
Waste from demolition of old buildings		1,000-1,500
Waste from mines and mineral industry		20,000
Mud from sewerage (15% of this correspond to dry substance)		1,500
Manure (10% is dry substance)		16,000
Straw, haulm etc.		10,000
Bark		1,200
Waste from food industry		500
Cars		150
Chemical waste etc.		450
Car tyres		70-80
	Total	57,900

Source: The RoR interim report on recycling

Appendix 5. List of Figures and Tables

Tables Page

Figures

Subject Index

1) *Page numbers within parantheses indicate that the* <u>*theme*</u> *has been dealt with on that page but no explicit mentioning of the word.*